Roman Bee Lore

By

L. E. Snelgrove

Roman Bee Lore - L. E. Snelgrove

Roman Bee Lore

Copyright © 2023 L. E. Snelgrove

All rights reserved. No part of this publication may be reproduced, stored in a retrieval system, transmitted in any form or by any means electronic, mechanical, including photocopying, recording or otherwise without prior consent of the copyright holders.

Roman Bee Lore was written by L. E. Snelgrove in 1922 but for some reason remained unpublished.

The manuscript was presented by the author's daughter to The Bee Research Association and is now published by Northern Bee Books in the interests of The Craft

ISBN 978-1-914934-57-5

Published by Northern Bee Books, 2023
Scout Bottom Farm
Mytholmroyd
Hebden Bridge HX7 5JS (UK)

Design and artwork by DM Design and Print

CONTENTS

Preface	V
Chapter 1. Introduction	1
Chapter 2. Importance of Apiculture amongst the Romans	12
Chapter 3. Natural History and Economy	17
Chapter 4. Management	78
Chapter 5. Honey	109
Chapter 6. Wax	123
Chapter 7. Pollen	133
Chapter 8. Propolis	136
Chapter 9. Honey Beverages	140
Chapter 10. Bee Flowers	147
Chapter 11. Curious Beliefs	157
Chapter 12. Didymus. De Apibus	166
Bibliography	168

ILLUSTRATIONS

		Page
1.	Hieroglyphic Bee	2
2.	Melissà, or Bee Goddess	4
3.	The Bees in Hive	25
4.	Brood Comb	31
5.	The Queen Bee	35
6.	Vergil's Battle of the Bees	39
7.	The Drone	41
8.	The Worker Bee	46
9.	A Large Swarm	53
10.	Swarming	60
11.	Cow-dung hive. South of France	64
12.	Cork hives. Spain	80
13.	Curious Hollow Tree Hive	82
14.	Earthenware and Mud Hives	84
16.	Roman Hives	86
17.	Watering - Place for Bees	91
18.	Ancient Beekeeping Apparatus	100
19.	Portion of Roman Wax Torch	128
20.	Roman Wax Tablet and Stylus	130
21.	Pollen-laden Bee	135
22.	"Oxen-born" Bees	160
23.	The Drone-fly (*Eristalis tenax*)	163
24.	An Old Exmoor Beekeeper	167

PREFACE

Until the end of the 18th century the didactic Greek and Latin writers on Agriculture were widely regarded as authoritative in their views and their writings, or later works based on them, commonly served as practical guides in husbandry. Modern developments in agricultural science and practice, however, have so largely detracted from their value that with one or two exceptions they are not now published in any language and it is exceedingly difficult to obtain copies of them either in the languages in which they were originally written or in translations.

In this paper I have collated and commented upon practically everything of importance in the old Latin writings relating to Bees and have, perhaps, rescued from oblivion a great deal of information respecting ancient beekeeping which should prove to be quite new to British readers and also, I hope, not uninteresting to many who are not engaged in the delightful pursuit of Apiculture.

In my treatment of what has proved to be a rather complicated subject I have endeavoured to show that the Romans not only possessed apiarian knowledge far more considerable than is commonly supposed and that in many respects they were in advance of rustic beekeepers of to-day, but also that the art of beekeeping remained practically stationary throughout mediaeval times.

In order to make my subject more intelligible to the lay reader I have given such information respecting Natural History and Modern Beekeeping as from time to time appeared to be necessary, and for the convenience of the examiner I have included in the text a large number of quotations, many of which, if this paper be published, will be omitted altogether or written as foot-notes.

Except in the cases of Varro and Vergil I know of no modern commentaries on the Latin writings on Apiculture. The old annotations available have proved of little value for the purpose of this paper, not so much because they are sometimes inaccurate, as because their scholarly authors frequently accepted without comment statements which were not technically true and, on account of lack of practical experience, omitted to interpret others so as to give them force and meaning in the light of the apicultural knowledge of their times. I have preferred, therefore, as a rule, to give my own comments and interpretations rather than to review those made by others.

Of the chief books studied I have quoted from the following texts: -

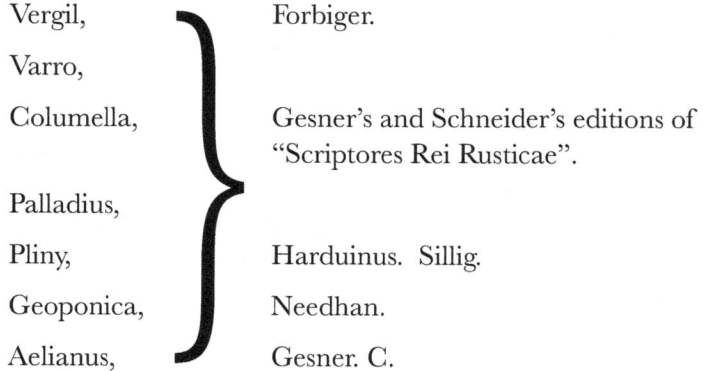

Editions of the other books used are specified in the Bibliography. Many of these have become very rare and I have been able to obtain them only with great difficulty. For help in this direction I am specially indebted to: -

(1) Mr T. W. Cowan, F.L.S., F.R.M.S. President of the British Beekeepers' Association, who kindly lent me a number of books from his comprehensive Beekeeping library.

(2) Professor Francis, Vice-Chancellor of the University of Bristol, who gave me permission to read in the University Library.

(3) The Bristol Central Library, which contains some fine folio editions of many of the less known Greek and Latin writers.

I have to acknowledge the courtesy of the following gentlemen who have kindly given me permission to use certain of the illustrations: -

(1) The Rev. G. H. Hewison, F.R.M.S. (Nos. 3, 4, 5, 6, 7, 21)

(2) Mr A. H. Meysey-Thompson (No. 13)

(3) Mr A Richards (No. 12)

(4) The Gresham Publishing Company (No. 23)

I am indebted also to Mr A. Smith, curator of the Greek and Roman Department, British Museum, for permission to take photographs for illustrations 1, 2, 14, 19, 20.

Preface

I have not considered it necessary to specify in the Bibliography much of the modern literature on Apiculture which has formed part of my general reading. This includes the best known British and American books on the subject as well as current Bee Journals published in England, America, France and Italy.

L. E. Snelgrove,   April 1922

Roman Bee Lore - L. E. Snelgrove

I
INTRODUCTION

ANTIQUITY OF APICULTURE

Of all the creatures domesticated by man the bee has remained, until quite modern times, the least understood and yet one of the commonest and most valued providers of his food. The mystery of its mode of generation, the supposed divine origin of its honey and its comparative immunity from being handled and observed within the walls of its home tended in ancient times to make it an object of curiosity, superstition and even reverential awe. It is not surprising therefore that the earliest writings of man contain references to bees and their products and that a large body of literature, with myth, superstition and fact curiously intermingled, has grown up with reference to apiculture.

The mutual adaptations of insects (especially bees) and flowers suggest that they have attained their present forms in the process of evolution *pari passu*, and that if this be so bees have an antiquity far greater than that of man. The presence of the remains of insectivora in pre-glacial formations of the earth's surface points to the same conclusion.

From the existence, in association with Neolithic remains, of traces of many phanerogamus plants, it has been inferred that bees, - insects specially adapted for the fertilisation of the flowers of these plants, - were common on the earth during the later Stone Age.

But if, as some suppose, the peculiar economy of the social bees is the result of evolution, then no stronger evidence of astonishing antiquity is needed than that presented by the organisation of a colony of bees, - consisting as it does of many thousands of females (the queen and the workers) together with a few hundred males (drones), the whole duty of procreation being confined to a single female but once mated. As no apparent developments in bee economy have occurred since the time of Aristotle, what countless ages must have elapsed since the time when the honey-bees, as may be assumed, lived alone or in tiny communities, every individual being capable of procreation.

The earliest records made by man were hieroglyphic in character. One of the oldest manuscripts of Egypt, "The Medical Papyrus" of the 14th Century B. C., contains a reference to honey which is prescribed as an important medicament.[1] In the hieroglyphics from the 4th dynasty onwards the bee is the symbol of Royalty – a sovereign reigning over an organised community. The figure of a bee is a component of several hieroglyphic combinations. Side by side with the lotos and placed before the name of a king, it was

---

1 Shuckard, p. 90.

the symbol for Lower Egypt, as distinguished from Upper Egypt denoted by the lotos. In other combinations it helped to symbolise a certain priestly order, and the ideas "substance", "serpent", "arrow", "bier" and "honey".[2]

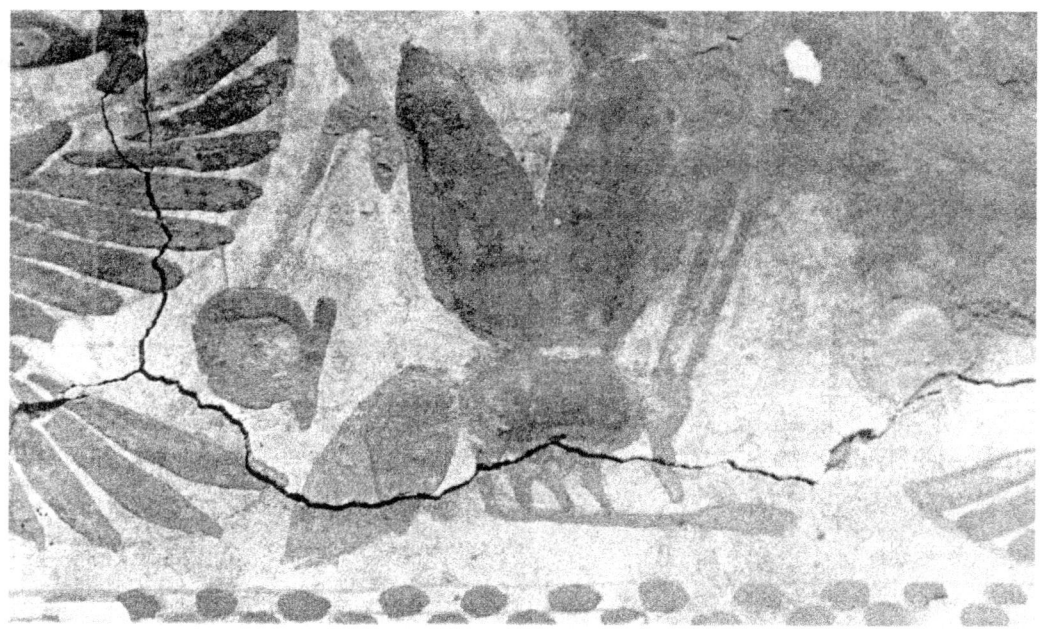

HIEROGLYPHIC BEE

Photographed from the coffin-lid of a priest of Amen-ra, Thebes.

XXII Dynasty, Circa 750 B.C.

British Museum

Shuckard refers to the Sanskrit word "ma", honey, with its derivatives "madhupa", honey-drinker and "madhukara", honey-maker and to the Chinese word "mih" or "mat", honey, as indicating the existence of a cognate word in the earliest (pre-Aryan) language and as implying some knowledge and perhaps culture of bees in the cradle of the human race.[3]

---

2 Morley, pp. 247-8.

3 Shuckard, p. 92.

MTHICAL AND OTHER REFERENCES TO BEES

Bees, honey and wax are frequently mentioned in connection with the earliest of the great nations. Homer, for example, (circa 850 B.C.), referring to the calling out of the Achaians to war likens them to thronging bees swarming from hollow rocks and settling amongst spring flowers.[4]

Isaiah (circa 750 B.C.) speaks of "the bee that is in the land of Assyria".[5]

Herodotus (B.C. 484-424) states that the Babylonians were accustomed to preserve dead bodies in honey[6] and that the Persians used wax for the same purpose.[7] This practice appears to have been fairly common. Diodorus Siculus referring to the Spartan King Agesilaus who died at Cyrene about 880 B.C. states that his body was taken home to Sparta embalmed in honey.

… "Corpus ejis melle conditum referetur Spartam".[8]

Plutarch, however, referring to the same incident, states that not having honey the Spartans used melted wax to preserve the body of the King.[9]

Classic myths include frequent references to bees. They were considered to be closely associated with and specially favoured by the Gods, whose food consisted of "nectar" and "ambrosia". One of the most interesting of these myths relates to the saving of the life of Zeus, the father of the Gods, by Cretan bees. The story goes that Cronos the father of Zeus had devoured all his children except Zeus. In order to save her last child his mother Rhea hid Zeus in a Cretan cave. Here he was nourished and his life saved, by the honey of bees dwelling in the cave and by the milk of the goat Amalthea. In return for this service Zeus became the special protector of bees and was supposed to have bestowed on them the following special rewards: -

(1) The privilege of wearing golden yellow bands (characteristic of Italian Bees),

(2) The ability to construct waxen cells and store honey therein, and so to exist in large families throughout the winter (as distinguished from the non-social bees, which, with the exception of their queens, die off at the approach of winter)

(3) The power to propagate their race without sexual intercourse and to live in highly organised communities.[10]

4	Iliad, II. 88, sqq.
5	Is. VII. 18.
6	Herod. I. 270.
7	Herod. I. 204.
8	Diod. Sic, XV. 93.
9	Plutarch, Lives, p. 432.
10	Edwardes, Lore. Intro. X. Vergil, Georg. IV. 153-5 & 197-209; Diod. Sic. V, 70.

Hyginus accounts for the origin of bees by relating the mythical belief that Jupiter transformed a beautiful woman named Melissa into a bee.[11] From her the bees received their name "μέλισσαι" or "μέλιτται" and the priestesses of certain goddesses propitious to the bees, - especially Cybele and Ceres, - were called "Melissae".

MELISSA, OR BEE GODDESS

Enlarged photograph of a pendant (electrum metal) used as a personal ornament. Taken from a tomb at Camirus, Rhodes. Date, circa 666 B.C. Now in the British Museum.

The figure shows a female head and bust, with outstretched arms and clenched hands, as well as conventional re-curved wings and the abdomen of a bee, the latter having eight segments instead of six. In the bottom corners are two conventional eight-petaled flowers.

---

11   Col. 9.2.3.

According to Ovid, the god Bacchus, a son of Zeus was the first to discover honey. "… a Baccho mella reperta ferunt".[12]

Travelling one day from Hebrus in Greece accompanied by Silenus and the Satyrs, he had reached Rhodope when the clashing of cymbals caused swarms of bees to follow. Bacchus secured them and having put them into a hollow tree subsequently secured the reward of some honey.

"Et inventi praemia mellis habet."[13]

Silenus having tasted its flavour sought for honeycombs in the woods. He found a stock of bees in a hollow tree, kept the knowledge to himself and seated on his ass approached the bees. Standing on the animal's back he attempted to take the honey, but multitudes of the bees stung his face and bald head. He fell headlong and at the same time received a kick from the ass. In response to his cries the Satyrs came to his aid, laughing at his swollen face. The God (Bacchus) laughed also, but showed Silenus how to use moistened earth as a remedy for the stings,

" … limum inducere monstrat

Hie paret monitis et linit ora luto …"[14]

Bacchus was reputed to enjoy honey, -

" … Melle pater fruitur",[15]

and it becomes, in association with wax, wine and cakes, an indispensable article in the offerings made at the great Bacchanalian feasts.

Amongst others of the Gods who were considered to take special interest in bees were:[16]

Pan, the God of woodland and country life in general, whose pipe was supposed to be an efficacious as the "tinnitus" in causing swarms to settle and to whom offerings of honey were specially acceptable; Cupid who dipped some of his darts in honey; Apollo, the sun God, and therefore guardian of bees, – "creatures of the sunshine"; Adonis, honoured with honey cakes and waxen fruit and flowers at his feasts; Aristaeus, who taught man the management of bees and was himself considered to be a metamorphosed swarm; Priapus the guardian of flocks and swarming bees; the goddesses Cybele, Ceres, Diana and Proserpine; and Mellonia the honey goddess of the Romans, - "Strong & powerful in regard to bees, caring for and guarding the sweetness of their honey".[17]

The most interesting mythical reference to wax relates to Daedalus, cunningest of Greek Artificers and builder of the famous labyrinth of Minos, King of Crete. Having

---

12   Fasti, III, 744.
13   Fasti, III, 744.
14   Fasti, III. 760.
15   Fasti, III. 762.
16   "M" B.B.J. 1916. pp. 66-7, & Moncrieff. Class. Myths, p.11.
17   Arnob. IV, 7.

offended his master, Daedalus was obliged to fly from the island with his son Icarus. The sea barred the way, but Daedalus made for himself and his son feathered wings which he fastened to their shoulders with wax. With these they flew towards Italy. The father arrived safely, but Icarus thoughtlessly flew high and too near the sun with the result that the wax melted, his wings became unfixed and he fell into the sea and was drowned near Icaria, the island named after him.

Mythical references to bees, honey and wax could be quoted in great number not only from Classical but also from Indian and Scandinavian mythology. None of them, however, points definitely to any organised practice of apiculture or actual management of bees. The bee is regarded as a mysterious creature in which the gods were specially interested; honey as a luxurious food, a valuable medicament and appropriate for libations in offerings to the gods; and wax as useful in crafts. The bees to which reference is made live in hollow rocks and trees. There is no mention of hives and from this we may infer that during the early periods in which these mythologies developed most of the honey and wax used was obtained from bees living in the wild state, as is the case today in many parts of Asia and Africa.

SCRIPTURAL REFERENCES

It is remarkable that the Scriptures contain no references to hives nor do they indicate that the Hebrews practised apiculture, - but references to bees and their products are numerous, e.g.: -

(1) To bees as vicious and dangerous: -
"The Amorites . . . chased you as bees do."[18]
"They compassed me about like bees."[19]

(2) To bees as being wild: -
"With honey out of the rock should I have satisfied thee."[20]

"And his meat was locusts and wild honey."[21]

(3) To honey as being good and delightful to the taste: -
"What is sweeter than honey?"[22]
"The judgments of the Lord ... sweeter also than honey and the honeycomb."[23]

18	Deut. I. 44.
19	Ps. 118. 12.
20	Ps. 81. 16.
21	Matt. 3. 4.
22	Jud. 14. 18.
23	Ps. 19. 9-10.

"Pleasant words are as a honeycomb, sweet to the soul and health to the bones."[24]
"My son, eat thou honey for it is good."[25]
"It was in my mouth as honey for sweetness."[26]

(4) To honey as a symbol of richness, - desirability: -
"A land flowing with milk and honey"[27]
"Thy lips, O my spouse, drop as the honeycomb; honey and milk are under thy tongue."[28]

(5) To the risks of eating too much honey: -
"Hast thou found honey? Eat so much as is sufficient for thee lest thou be filled therewith and vomit it."[29]

(6) "It is not good to eat much honey: so for men to search their own glory is not glory."[30]

(7) To the melting of wax: -
"My heart is like wax; it is melted in the midst of my bowels."[31]
"As wax melteth before the fire, so let the wicked perish at the presence of God."[32]

In the Old Testament honey is several times mentioned as an article of barter or gift, but, unlike the Greeks and Romans, the Hebrews did not use it in their burnt offerings. - "Ye shall burn no leaven nor any honey in any offering of the Lord made by fire."[33]

There is no evidence that they had domesticated the bee notwithstanding their appreciation of honey, and it is most likely that they were content to despoil the nests of wild bees with which the hollow rocks of Palestine abound even to this day.

24	Prov. 16. 24.
25	Prov. 24. 13.
26	Ezek. 33.
27	Ex. 3, 8.
28	S. of Sol. 4, 11.
29	Prov. 25, 16.
30	Prov. 25.27
31	Ps. 22. 14.
32	Ps. 68. 2.
33	Lev. 2. 11.

APICULTURE AMONGST THE GREEKS

The period when man first began to domesticate bees by hiving swarms is unknown. Possibly individuals in all ages had succeeded in doing this, but the credit of the earliest systematic bee management is probably attributable to the ancient Egyptians. The earliest known representation of a hive is delineated by Wilkinson who found it on a very ancient tomb of Thebes in Egypt.[34] The apicultural knowledge and skill possessed by the ancient Egyptians were almost certainly absorbed and perpetuated by the Greeks.

The exquisite honey of Mount Hymettus became world famous and naturally conduced to the development of Apiculture in Greece. It is not surprising therefore that Aristotle, the greatest of ancient writers on Natural History, devotes the whole of one of his books to a description of the nature and management of bees. The Latin writers on agriculture were largely indebted for their material to Aristotle and possibly also to the lost Μελισσουργικα of Nicander. These writers in their time were doubtless similarly indebted to their predecessors.

The knowledge of bees possessed by the Romans was therefore largely derived from the Greeks, who in turn had doubtless learnt much from the ancient Egyptians. There is no reason to suppose that any other peoples contributed in any appreciable degree to the knowledge of Apiculture as understood and practised by the Romans during the period to which this paper principally applies, viz., B.C.100 – A.D.300.

BEES FAMILIAR TO THE ROMANS

The natural Order of the Hymenoptera, - insects distinguished by the possession of four membraneous gauzy wings, - includes more than two thousand distinct species of bees. By far the most important of these are the Apidae, or long-tongued bees, which include the varieties of the honey-bee (*Apis Mellifica*) domesticated by man and which, until the advent of sugar, provided him with his only important sweet.

The varieties of the honey-bee native to Europe and the shores of the Mediterranean and which therefore were familiar the Romans are: -

(1) *Apes Mellifica*, variously known as the common black, brown, or German bee. This is indigenous to the whole of Europe with the exception of countries contiguous to the Mediterranean and Adriatic seas. Until the 19th Century this was the only hive-bee cultivated in the British Isles and it is still the commonest bee of Northern Europe, Africa and Western Asia. It was introduced by settlers and spread rapidly in America and Australia, but in these countries today it

---

34   Shuckard, p. 90.

is gradually being displaced by the Italian bee. English, German, Dutch and Carniolan bees are well-known but slightly differing examples of this race. *Apis Mellifica* is distinguished by the dark brown and almost black colour of its body, relieved to a greater or less extent by light grey pubescence on its thorax and between its abdominal segments.

(2) *Apis Mellifica*, var. Ligustica, native to Italy. These bees are distinguished by more or less bright yellow bands or rings on the three first abdominal segments which give them a beautiful appearance. They are undoubtedly the bees referred to by Vergil as "Ardentes auro et puribus lita corpora guttis".[35] They inhabit the whole of Italy, the litoral of the Adriatic Sea and parts of Greece and Turkey. They have kept pure and unrelated to the German and Carniolan black bees on account of the natural barrier formed by the Alps.[36]

(3) *Apis Mellifica*, var. Fasciata, or the Egyptian bee, with its near relatives the Palestine or Syrian and the Cyprian bees. These are rather smaller than the Italians, possess similar abdominal markings and, in addition, a bright yellow patch ("scutellum") on the back of the thorax. They are uncertain in temper, impatient of ordinary methods of management and have not proved either popular or successful in other countries where they have been tried.[37]

The Romans were doubtless familiar also with the numerous species of the non-social (wild) bees inhabiting Italy and neighbouring countries but references to them are rare. Wasps, however, are frequently mentioned, especially as redoubtable enemies of the honey-bees.

LATIN WRITERS ON BEES

Most of the classical Roman writers make isolated references to bees, honey and wax; a few however deal systematically with Apiculture. Of these the earliest was: -
1. VARRO, who lived in the first Century B.C., an educated and much travelled man and a prolific writer. Varro composed a treatise on Agriculture – "De Re Rustica" which included a long chapter on Bees. For materials for this chapter he was largely indebted to Aristotle's "Historia Animalium" written about 300 years earlier. The "De Re Rustica" is the only complete work of Varro's which remains. It was preserved in a collection of Agricultural treatises commonly

---

35 Vergil, Georg. 4, 99.
36 Langstroth, p. 292 et seq.
37 Ibid.

entitled "Scriptores Rei Rusticae", all the editions of which have descended from a single manuscript preserved at Florence until the 17th Century. The text has suffered much through successive copyings in manuscript.

Contemporary with Varro was the poet

2. VERGIL, who devoted the whole of his fourth Georgic to Bees. This is considered by many to be the most beautiful didactic poem ever written. Throughout it the poet extols toil and industry of which the bee is an exemplar. As Connington says: -

"The pervading atmosphere of the book is one of labour from beginning to end".[38]

Vergil appears to have borrowed considerably from both Aristotle and Varro for his materials, but he is never uninteresting and the poem contains some delightful descriptions of the life and duties of the bees. Although Vergil was undoubtedly a close observer and lover of bees and possibly kept them in his garden at Mantua, there is nothing in his writings to indicate that he was a practical bee-man as some have supposed. The fourth Georgic however remained the best-known work on Apiculture throughout the Middle Ages and Edwards, writing in 1908, even goes so far as to say that "for the beginner in Apiculture today no wiser choice than this fourth book of the Georgics could be made" – which is, of course, a charming exaggeration.

3. PLINY THE ELDER, who perished at the eruption of Vesuvius in the 79 A.D., wrote an encyclopaedic Natural History in thirty-seven books, of which considerable portions of the eleventh and twenty-first are devoted to bees and their products. He repeats much of the information given by Aristotle and Varro and in addition writes fully respecting the uses of honey, wax and propolis, especially as ingredients of medicaments. He is serious and often critical of others' opinions, but he mingles with a mass of useful detailed information many obvious absurdities.

He refers to two earlier writers on bees, whose books have disappeared, - Aristomachus of Soli, who made Apiculture his sole occupation, and Phiscus of Thasos who spent his whole life in the contemplation of bees.[39]

4. LUCIUS COLUMELLA, one of the "Scriptores Rei Rusticae", a Spaniard who flourished about the middle of the first Century A. D., is the most important and accurate of the ancient writers on Apiculture. He devotes the whole of

---

38 Connington, Intro. to Georgics.
39 Pliny, H.N. XI, 9.

the ninth book of his "De Re Rustica" to bees. He is an admirer of Vergil, from whose works he quotes, and he refers to Hyginus, Celsus Nicander and the Carthaginian Mago – previous writers on Apiculture – whose works have since been lost. Columella is eminently practical. There are evidences in his writing that he kept and managed bees himself – a statement it would be difficult to make about the other Latin writers on the subject. He leaves speculative considerations to other writers and confines himself to what he considers useful to practical agriculturalists.[40]

5. PALLADIUS, who probably lived in the early part of the 3rd Century A.D., was another of the "Scriptores Rei Rusticae." He wrote a treatise on Agriculture in the form of a calendar giving instructions for the work of each month. His writings include twelve chapters on Bees, much of the material for these being taken almost literally from Columella.

6. The GEOPONICA is a compilation of the writings of ancient Greek and Roman writers on Agriculture, probably made by one Cassianus Bassus in the 10th Century A.D., for the Emperor Constantine Porphyrogeneta. Little or nothing is known of the writers on bees in this collection or of the times when they lived. They are: -

(1) FLORENTINUS, who probably lived about 218 A.D. He wrote a long chapter "De apibus, et quomodo ex bove generentur."

(2) DIDYMUS, who wrote
(a) "De apibus"
(b) "Ut apes non fugiant."
(c) "Quando apes vindemiare oporteat."

(3) PAXAMUS, who gives directions "Ut vindemiator mellis non pungatur."

(4) DIOPHANES, who writes "De melle et ejus cura."

(5) LEONTINUS, who shows how one may ensure "ut neque alvearia, neque agri . . . (etc) ullo incantamente afficiantur."

(6) DEMOCRITUS, who gives directions "Ad perimendos fucos."

(7) AELIANUS (circa 200 A.D.) a Roman, who wrote in Greek, devotes several chapters of his work on Natural History to bees. He exaggerates the admirable qualities of the bee but says little that is not mentioned by one or more of the writers who preceded him. From certain statements he makes it seems unlikely that he had any practical experience of apiculture.

---

40 Col. 9. 2. 5.

2
ROMAN BEE LORE

IMPORTANCE OF APICULTURE AMONGST THE ROMANS

The high value placed on honey as the most important available sweet and as a medicament, the ease with which delectable fermented drinks were made from it, the innumerable uses of bees' wax and the prescription of these substances for religious sacrifices all tended to promote the pursuit of Apiculture amongst the Greeks and Romans. All the writers "De Re Rustica" considered it an essential feature of good agriculture. Pliny states that hives properly managed provide considerable profit to their owners,[1] and Varro quotes two striking examples of the profits made by beekeepers:[2]

Seins received a yearly rental of 5000 lbs of honey for letting out his hives ("quotannis quinis millibus pondo mellis").[3] Assuming that he extracted as much as half the total average crop as his share, this would represent at least 500 beehives and indicates that he was a bee-farmer on a very extensive scale.

Two brothers named Veianius who had served under Varro in Spain having inherited about an acre of land from their father surrounded it with beehives and planted it with thyme and other honey-bearing flowers with the result that they never took less than 10,000 sesterces (over £80) profit each year. They managed this by selling their honey only when the market was favourable, - "suo tempore potius quam celerius".[4] These brothers set an example to most present-day apiarists, who rush their honey to market as soon as it is gathered in the summer time and consequently have to accept lower prices than they could obtain in the winter or in seasons when poor crops are obtained.

As an example of a professional apiarist Pliny refers to Aristomachus who did nothing but keep bees for fifty-eight years.[5]

Vergil portrays a delightful example of the amateur beekeeper of his day.[6] His Corycian, an old man possessing a few acres of poor land not good enough to be

---

1	Pliny, H.N. XXI, 41.
2	Varro, R.R. III, 16 (10-11).
3	Varro, R.R. III. 16 (10-11).
4	Varro, R.R. III, 16. 11.
5	Pliny, H.N. XI. 9.
6	Vergil, Georg. IV, 125-146.

ploughed, - "nec fertilis illa juvencis", and unsuitable for good crops or even pasture, by diligent work had succeeded in making his small-holding provide him with many unbought dainties, - "dapibus inemptis," with which he loaded his table. He was the first in season to pluck flowers and fruit and to be blest by early swarms and foaming honey pressed from the combs

" … examine multo,

Primus abundare et spumantia cogere pressis

Mella favis".[7]

His fruit trees were productive, the fruit promised at blossoming time arriving at maturity in the autumn, -

"Quotaque in flore novo pomis se fertilis arbos
induerat, totidem auctumno matura tenebat,"

- an unconscious reference by Vergil to the fact, now commonly recognised, that crops of fruit largely depend on the fertilisation of the blossoms by means of bees. In contentment of mind this simple countryman "equalled the wealth of Kings," –

"Regum aequabat opes animo".

A Roman farm was not complete without its apiary. Both Varro and Columella give detailed advice as to the choice of a site for the hives. Some people kept them in the porticos of their houses.[8] Others placed them on specially built brick or stone embankments near to the farm-house where they would be safe from thieves. Others again kept them in sheltered valleys near the farms where they could readily be supervised. Frequently an apiary was so important as to need the whole attention of a custodian called the "mellarius" or "apiarius" who lived in a specially provided house ("tugurium") replete with tools and prepared empty hives.[9]

Immense quantities of honey and wax were imported into Italy, partly in the form of tributes from subject states. The most prized came from Attica, Sicily and the island of Calydna off the coast of Caria.[10] Crete, Cyprus and the North of Africa also provided honey of superior quality.[11] Inferior honey, - "mel asperimmum," came from Sardinia and Corsica. Diodorus Siculus tells us that the Carthaginians extorted a tribute of refined honey and wax from Corsica, which abounded in these products, -

---

7	Ibid.
8	Varro, R.R. III, 16. 15.
9	Colum. R.R. IX, 5. 3.
10	Pliny, H.N. XI, 13.
11	Pliny, H.N. XI, 13.

"Corsicae urbes in potestate habent tributa ab incolis exigebant refinam ceram, mel, quorum largus ibi proventus est."[12]

From the same writer we learn that when the island was conquered for Rome by the praetor Pinarius, 181 B.C., it was subjected to an annual tribute of 100,000 pounds of wax – subsequently increased to 200,000 pounds.[13] At a low estimate such an amount of refined wax must have corresponded to at least as many stocks of bees and perhaps the enormous total of 2,000,000 pounds of honey! Diodorus states, however, that swarms were commonly found in the hollow trees in the hilly districts of the island and that they became the property of those who found them, -

"Namquae in cavis montium arboribus inveniumtur mella, citra controversiam illorum sunt qui invenerunt."[14]

In any case the duty of procuring so much wax must have involved an enormous industry.

From Pliny we learn that the Sanni, a people of Pontus, also paid a tribute of wax to the Romans although they did not sell the honey produced with it on account of its poisonous properties.[15]

It is a commonly recognised fact today that a particular locality may be overstocked with bees and the unprofitableness of such over-crowding was not unknown to the Greeks and Romans. According to Plutarch, Solon (circa 598 B.C.), when abolishing the severe laws of Draco, retained one relating to bees, ordaining that "if anyone would raise stocks of bees he was to place them about 300 feet from those already raised by another."[16]

Celsus, cited by Columella, points out that some localities provide adequate pasturage for bees in the spring and summer but not in the autumn and vice versa, - few places being satisfactory throughout these seasons. He therefore recommends the important practice commonly known today as migratory beekeeping:

" ... vernis pastiones absumptis, in ea loco transferri quae serotinus floribus thymi, et origani, thymbraeque benignius apes alere possunt",

and states that it is common, on the approach of autumn, to transport hives for late pasturage, -

---

12	Diod. Sic. V, 13.
13	Diod. Sic. V, 13.
14	Diod. Sic. V, 14.
15	Pliny, H.N. XXI, 45.
16	Plutarch, Lives, p. 68.

(1) From the districts of Achaia into Attica

(2) From Euboca and the Cyclades into the island of Scyros

(3) From parts of Sicily to Hybla.[17]

Pliny states that the inhabitants of Hostilia, a village on the banks of the Po, when bee pasturage fails them, take their hives upstream in boats by night and after liberating the bees and changing the position of the boats from day to day at last arrive home, the crop of honey being indicated by the depth to which the boats have sunk in the water.[18] Bees commonly fly two miles or more in all directions for their pasturage and as the five miles, - "quina milia passuum," - through which the boats were moved made so much difference, the beekeeping of Hostilia was probably conducted on a comparatively intensive scale. From Pliny also we learn that in Spain hives were moved to new pasturage on the backs of mules.[19]

Migratory beekeeping has been practised with advantage from the earliest times. The ancient Egyptians had floating apiaries, which, after being taken upstream on the Nile moved slowly down "following the blossoming of the plants along the banks as the annual inundation receded."[20] Rotter states that these floating apiaries passed from Upper Egypt down to the sea between December and the end of February.[21]

Modern apiarists in many parts of the world still move their hives "to the heather" after the summer season. In the north of England and in Scotland this is done by means of carts, - sometimes even by rail. In mild autumns large harvests of dark-coloured thick honey are obtained, chiefly from the common ling (Erica Vulgaris). In Italy itself, according to Benussi-Bossi and Sartori, "L'apicoltura nomade" is little known and practised today (1901), except in the provinces of the Udine, Sicily and Sardinia. These writers point out, however, the rich possibilities of moving hives from the Alpine districts to the plains in the spring and vice versa in the autumn, -

> "Gli apicoltori alpini dovrebbero transportare i loro alueari al piano, dove specialmente domina il gelso e si coltiva il ravettone. In Maggio, finita, al piano, la stagione del trifoglio o delle acacie, si smela il raccolto fatto e di notte si traslocano alle prealpi, che danno un raccolto un po' ritardato ma ricco in guigno e juglio, per ritornare in

---

17   Colum. R.R. IX, 14. 19.
18   Pliny, H.N. XXI, 43.
19   Pliny, H.N. XXI, 43.
20   Morley, p. 252.
21   Bee World, April 1921, p. 81.

> agosto e settembre nei luoghi dove fiorisce l'erica, si coltiva il grano Saraceno e si offre spontanea e ricchissima la raccolta del brugo."[22]

We have very little information respecting trade in bee products in ancient times. Of the price of honey, we know nothing. Varro says that propolis, used by doctors in the making of plasters ("emplastris") was dearer than honey when sold in the Via Sacra.[23] Wax was cheap, - probably on account of the tributes from Corsica and Pontus. Columella speaks of it as not to be neglected although of little pecuniary value, "Cerae fructus, quamvis aeris exigui, non tamen omittendus est."[24]

From the foregoing considerations we may conclude that Apiculture was extensively practised in the Roman dominions. It was a profitable pursuit for people sufficiently skilled to take honey and wax from the wild swarms found in hollow rocks and trees; it augmented the revenues of the small-holder and farmer and provided a much-desired delicacy for their tables; and it was sometimes conducted on a large scale either by professional apiarists or by the inhabitants of good honey-bearing districts, especially where "migratory" beekeeping proved profitable.

---

22 L'arte de coltivare le Api, p. 227.
23 Varro, R.R. III, 16. 23.
24 Colum. R.R. IX, 16. 1.

3
NATURAL HISTORY & ECONOMY

ANATOMY AND PHYSIOLOGY

Being without the advantage of magnifying glasses the ancients had comparatively little accurate knowledge of the structure of the bee and the functions of its various organs, and many of their conclusions, based on casual and superficial observation, were inevitably fallacious. Aristotle devotes several paragraphs to the anatomy of insects, and Pliny, the only one of the Latin writers who deals with it to any extent, follows him closely.

Aristotle places the bee amongst the "Insects", animals with divided or insected bodies,[1] and he later includes them in the nine species the make wax.[2]

Pliny describes insects, including bees, as animals whose bodies appear to be almost separated by constrictions at the neck and waist, - the head, thorax and abdomen being held together by little tubes, -

"tenui modo fistula cohaerentia",[3]

a distinction still applied in the modern classification of animals.

Instead of the bony framework possessed by most other animals, an insect has an "exoskeleton" consisting of a tough, elastic, horny substance known as Chitine. In the case of the bee this is arranged in segments which appear as rings encircling the body and which form an attachment for the support of the internal organs and muscles. In the thorax the segments are more or less firmly bound together but in the abdomen they are joined by a soft skin under-folded so as to allow them to telescope one into the other, thus permitting of great flexibility and power of extension of the lower part of the body. Pliny in describing this arrangement, - "imbricatis flexilibus vertebris", - observes that nowhere else is the workmanship of Nature more remarkable.[4]

Although insects have no lungs their bodies are completely aerated by means of a

---

1	Arist. H.A. IV, 7.
2	Arist. H.A. IX, 40.
3	Pliny, H.N. XI, 1.
4	Pliny, H.N. XI, 1.

wonderful branching system of fine tubes (tracheae) which communicate with the outer air through ten pairs of minute openings (spiracles) arranged along the sides of the body, eight pairs being located on the abdomen and two pairs on the thorax. Both the tracheae and the spiracles are so tiny that they necessarily escaped the notice of ancient observers who concluded that insects were without lungs and did not breathe. And so Pliny says, -

"Insecta multi negarunt spirare,"[5]

and Aelianus, -

"Insecta, ut vespae, apes; haec etiam pulmonibus carere dicuntur".[6]

They were considered to live as plants and trees, but Pliny himself does not accept this view without reserve and does not see why they should not be able to breathe notwithstanding the absence of lungs; unconsciously he gives two illustrations which support his theory, viz.,

(1) If oil be applied to their bodies it kills them (probably because it blocks the spiracles) -

"Oleo quidem non apes tantum, sed omnia insecta exanimantur, praecipue si capite uncto in sole ponantur."[7]

(2) They can be killed by their own honey if this is smeared on their abdomens (where it would inevitably choke the spiracles), -

"Nocent et sua mella ipsis inlitaeque ab adversa parte moriuntur."[8]

It was believed also that bees were without heart or blood, although Pliny speaks of a liquid contained in their bodies which he considers is the equivalent of blood.[9] The tube-like organ which extends through the dorsal region of the abdomen and which by rhythmic contractions keeps this fluid – which is really the blood – in motion and which therefore fulfils the function of a heart, was unknown to Aristotle and the Roman writers.

Bees were credited with a voice, the murmur which they were considered to make when diseased or when about to swarm being attributed to the vibration of the wings. We now know that in addition to the hum produced by the wings, bees are able to produce other sounds, probably by means of vocal apparatus in certain of the spiracles.

---

5   Pliny, H.N. XI, 3.
6   Aelian. XI, 37.
7   Pliny, H.N. XI, 19.
8   Pliny, H.N. XI, 19.
9   Pliny, H.N. XI, 3.

Natural History & Economy

None of the Latin authors gives detailed descriptions of the limbs, but Pliny, Columella and Palladius, in describing the "rulers" (queen bees), refer to differences in the lengths of wings and legs, in pubescence and in markings.[10] Pliny also attempts to describe the manner in which the hind legs are loaded with pollen ("flores") from the flowers. He says that this is placed on the inner sides of the thighs, - ("femina"), which are specially rough for the purpose, by means of the front feet, which in turn receives it from the mouth.[11] This is not true of the pollen however. It is not gathered by the mouth but is collected from the hairs of the body by special combs on the thighs of the hind legs.

His description could more appropriately be applied to the collection of Propolis, which, gathered from tree-buds by the jaws (mandibles), is passed by way of the feet until it is safely loaded on the thighs of the hind legs.

Following Aristotle, Pliny attributes all the senses, - "tot sensus" to insects, and speaking of the gnat he demands, with unfeigned admiration, - ("quam inextricabilis perfectio"!) – how Nature can have placed the organs of sight (visum), of taste (gustatum) and of smell (odoratum) in so tiny a creature.[12] Of insects in general he says that they possess the senses of sight, taste, smell, touch (tactum) and in a few cases, hearing, - "pauea et auditum".[13]

Aristotle denies that bees can hear[14] although he considers that they appreciate the noise made by the beekeepers to induce swarms to settle.[15] Most modern writers on bees consider that they can hear, but as Snodgrass says[16] there is no direct proof of this and the auditory organs are not definitely identified.

Pliny's limitation respecting the sense of hearing is interesting. The sense of SIGHT was evident from the possession of eyes; of TASTE, from the presence of the tongue, the obvious pleasure with which food, such as honey, was eaten, and the rejection of unnatural and unsavoury foods; of SMELL from the ease with which agreeably scented foods, such as honey, even though hidden from view, were detected and the abhorrence of disagreeable odours; and of TOUCH, from the apparent use of the antennae as feelers and extreme sensitiveness of insects to contact with external objects; but in the case of most insects there was no evidence of their ability to HEAR. The supposed

---

10 Pliny, H.N. XI, 16; Colum. R.R. X, 1; Pallad. R.R. VII, 7.
11 Pliny, H.N. XI, 10.
12 Pliny, H.N. XI, 2.
13 Pliny, H.N. XI, 4.
14 Arist. Metaphys. I, 1
15 Arist. H.A. IX, 40.
16 Snodgrass, Anatomy. P. 39.

efficacy of the "tinnitus", - a noise made to cause a swarm of bees to settle, - was however considered to prove that these insects could hear, -

"Gaudent plausu atque tinnitu aeris eoque convocantur, quo manifestum est auditus quoque inesse sensum"[17]

Even today, however, we know very little about the senses of bees. Special structures in the antennae are considered to be organs of touch and smell and certain others may be auditory. The evidence that bees can hear, however, is indirect only. It seems reasonable to suppose that they can hear one another's voices and also the piping of their queens which happens at swarming time. On the other hand, they appear to be entirely indifferent to loud noises in their immediate vicinity as demonstrated by Lubbock in his experiments on Ants, Bees and Wasps. Lubbock states, however, that Will, who repeated his experiments, found that the insects were excited and extended their antennae, when their own sounds were imitated by means of a file and a quill.[18]

It may be that bees are susceptible only to certain sounds which intimately concern them and which their auditory organs are tuned to receive. Cheshire regards Lubbock's experiments as inconclusive and says, -

"Tuning-forks, whistles and violins emit no sounds to which any instinct of these creatures could respond. Should some alien being watch humanity during a thunder storm he might quite similarly decide that thunder was to us inaudible. Clap might follow clap without securing any external sign of recognition; yet let a little child with tiny voice but shriek for help and all would at once be awakened to activity. So, with the bee; sounds appealing to its instincts meet with immediate response while others evoke no wasted emotion."[19]

STINGS

That bees are provided with a sting which can be used as a most effective weapon of defence is well known even to those who have not been privileged to experience its effects, and it is certain that it received the most serious attention of the earliest naturalists who ventured to study the habits of bees. It is not surprising, therefore, that Aristotle speaks with wonderful accuracy on the subject. He knew that the worker bees were provided with stings, that the drones were not, and that the rulers or "Kings", although possessing

---

17   Pliny, H.N. XI, 20.
18   Lubbock, Senses of Animals, p. 96.
19   Cheshire, Bees & Beekeeping. Vol. I. p.107.

them, did not use them – at least on human beings.[20] He goes on to say that a bee dies after having stung because it cannot withdraw its sting from the wound without self-mutilation; but if the person stung gently presses the sting out of the wound the bee continues to live.[21] All this is good natural history today and it was repeated with variations and not quite so accurately, by later writers on apiculture. Pliny, for example, states that nature has given a sting (aculeum) to the bee and that this is affixed to the abdomen; that some people considered that after a single application of this sting ("unum ictum hoc infixo") the bee soon afterwards ("statim") died; but that others believed that this happened only if the sting were so far inserted that the poor bee was obliged to leave behind not only the sting but a part of its intestines as well.

"Alique non nisi in tantum adacto ut intestini quippiam sequatur".[22]

The sting of a bee is composed of two sharp darts, not more than 1/400 inch in diameter, each of which is provided with a number of retrorse teeth or barbs, somewhat like those of a fish-hook. These darts slide along another spear-like instrument, known as the sheath, which itself is wonderfully sharp and provided with barbs. The sheath and darts together make an astonishingly fine spear-like weapon which, when once inserted in the flesh of any animal can be withdrawn only with great difficulty. In the act of stinging the darts are inserted to a depth of not more than 1/12 inch. The great pain experienced is not due to the puncture but to a poisonous liquid which flows down between the sheath and darts, as the latter move, and enters the wound. This poison is secreted and stored in a small bag, - the poison-sac, connected with the sting and situated within the posterior segment of the abdomen. When the stinging bee is violently pushed away by its victim this poison-sac is torn away from the bee's body and remains attached to the sting which holds fast to the wound on account of the barbed darts. It is the poison sac and its ducts to which Pliny refers as part of the intestines and it is the mutilation caused by the violent withdrawal of these from the bee's body that causes its death a few hours later. If left to itself a stinging bee will often succeed in withdrawing its sting from the wound by dint of repeatedly turning round it as on a pivot.

A stinging bee affords an example of implacable fury. In its effort to injure its foe it is oblivious to everything else, and as Vergil says, finally lays down its life in the wound; -

"Illis ira modum supra est, laesaeque venenum

Morsibus inspirant, et spicula caeca relinquunt

---

20 Arist. H.A. V, 21.
21 Arist. H.A. IX, 40.
22 Pliny, H.N. XI, 18.

Adfixae venis, animasque in volnere ponunt."[23]

Pliny expresses a difference of opinion amongst authors as to whether the "King" bee was provided with a sting, or whether he was armed only with majesty; or whether, having a sting, he was denied its use, for certain it was that he made no use of it and yet secured the wonderful obedience of his people.[24] The fact is that the queen bee (known to the ancients as a King), is provided with a sting which she uses only when contending with a rival queen, - a very rare occurrence. The early beekeepers must have handled queens (a feat of which many modern beekeepers cannot boast), before they discovered that they would not sting human beings.

Palladius speaks of the "Kings" as not using their stings, -

"… ventre, quo tamen non utuntur in vulnos".[25]

and Aelianus says, -

"… nunquam tamen eo neque contra hominem neque contra apes uti".[26]

Seneca in his book on "Clemency" affirms that the "King" has no sting, for Nature has refused him a weapon and left his wrath disarmed, -

"Natura … telumque detraxit et iram ejus inermem reliquit".[27]

The early naturalists did not make the mistake, - so common to many people of today, - of believing that bees and wasps are always ready to attack inoffensive persons. Thus Aristotle rightly says that bees will fight only near their hives and not away from them.[28] Varro describes them as, harmless, - "minime malefica", not destroying any man's work, although not slow to resist any who may attempt to disturb their own.[29] Didymus expresses the same truth in the following passage, -

"Neque vero aliorum labores devastat, resistit tamen fortissime iis, qui ipsiis labores destruere aggrediuntur".[30]

The modern beekeeper relies on two things to enable him to handle his bees without provoking them to sting. First it is essential that they be well fed and secondly all movements near the hive must be slowly and gently carried out. They are wrathful when

---

23   Vergil, Georg. IV, 236-8.
24   Pliny, H.N. XI. 17.
25   Pallad. VII, Tit. 7.
26   Aelian. I, 60.
27   Seneca, De Clementia I, 19.
28   Arist. H.A. IX. 40.
29   Varro, R.R. III, 16. 7.
30   Geop. XV, 3.

hungry and violent movements rouse them to fury. After the administration of smoke, they are good-tempered because in their alarm they feed liberally on their stores of honey. Both Greeks and Romans used smoke to pacify their bees when taking honey, but there is no indication in their writings that they knew why the smoke was effective. The chief motive in applying it was to drive the bees from their combs. Columella says that the wrath of bees is readily appeased on the intervention of those who care for the hives, but he goes on to state a fallacy which is still very prevalent, namely, that the oftener the bees are handled the sooner they become tractable, -

"nam cum saepius tractantur celerius mansuescunt."[31]

There is no reason for supposing that bees ever recognise persons. In the summer months individual bees do not live much longer than six weeks, and therefore have no opportunities for distinguishing one person from another, even if that were natural to them. Indeed, it is safer to say that the less they are interfered with the better-tempered they are.

Nothing provokes bees to sting more than the ridiculous gestures of some people who, the moment they are threatened, begin to beat the bees off wildly with their hands. Florentinus aptly refers to this when explaining why people are attacked, -

"Cum vero homines accendentes moleste ferant, ipsosque impetant, … saeviores inruunt".[32]

Several liquids have been compounded in modern times to serve as "apifuges". When rubbed on the hands they were supposed to discourage bees from stinging. The basis of most of them was Oil of Wintergreen. They have now gone out of favour, however, and are seldom heard of. The idea of an apifuge was by no means a modern one. Paxamus prescribes the following mixture to be rubbed on the face and bare parts of the body: -

"To the roasted flour of fenugreek add the juice of the wild mallow and some oil; make into a paste of the thickness of honey."[33]

It is quite unlikely that this mixture was of any use and Paxamus does not express an opinion. The truth is that no effective unguent has been found to serve as an apifuge and experience has shown that if bees are properly prepared and gently handled, no protection, at least for the hands, is necessary.

The number of popular remedies for the effects of bee stings is legion, most of them being quite useless. The poison injected by a sting is immediately carried away by the

---

31   Colum. R.R. IX, 3.
32   Geop. XV. 2.
33   Geop. XV. 6.

circulating blood, and local applications of remedies, unless immediate, have no value. The modern practice is to apply an alkali such as soda or weak ammonia immediately after the sting has been removed, on the assumption that the Formic Acid, the principal constituent of the bee poison, may be neutralised and rendered less effective. Pliny recommends several remedies, viz.,

(1) Apply the juice of the leaves of the mallow or of ivy.[34]

(2) Drink some pure wine (merum).[35]

(3) Apply melissophyllum (balm).[36]

(4) Apply vinegar (acetum), to allay the itching which often results from a sting.[37]

Only the last named of these would be likely to have a beneficial effect. Vinegar is commonly used today as a simple homely remedy for the irritation produced by bites and stings of insects.

KINDS OF BEES

A colony of bees comprises:

(1) The "queen", which lays eggs in the cells of the comb and is the mother of all the others.

(2) A few hundred "drones", the male bees which are reared during the summer months and exist only for the fertilisation of queens. The occasion for this is very rare, as a queen is ordinarily fertilised only once in her life, which, in a state of nature, lasts four or five years. The drones do no work of any kind are without stings. When the honey flow ceases at the end of summer they are expelled and starved to death by the "worker" bees.

---

[34] Pliny, H.N. XXI, 45.
[35] Pliny, H.N. XXIII, 23.
[36] Pliny, H.N. XXI, 86.
[37] Pliny, H.N. XXIII, 27.

Natural History & Economy

THE BEES IN A HIVE

In the centre, the Queen; on the left, a Worker; on the right, a Drone.

(3) From ten to fifty thousand or more "worker" bees, who form the main body of the hive population. They are females, but on account of certain limitations in their diet when in the larval stage they are not so large as the queen and their genital organs are insufficiently developed to permit of their being fertilised and producing eggs. From the flowers they gather honey and pollen for food, and from tree buds a gummy substance known as propolis with which to seal crevices in their dwellings. They do all the work of the hive, providing the heat necessary for the hatching of the eggs, feeding the larvae, queen and drones with pre-digested food, and producing the wax with which they build their combs. The length of their life is roughly proportionate to the amount of work they are called upon to do. In the summer it varies from six to twelve weeks; from autumn to spring it may extend to as much as six months. The queen may lay eggs at the rate of 3,000 a day in summer, but this number gradually diminishes towards the winter, during a brief period of which

breeding is discontinued. It recommences in January or February, steadily increasing until in May or June the small winter population of a few thousands has increased enormously, - only to dwindle again as winter approaches.

The early writers are all more or less confused in their descriptions of the different sorts of bees. Vergil, in agreement with Aristotle and Varro, expresses a preference for what must be the red or golden-banded Italian bees as distinguished from those which are dark in colour, -

"Elucent aliae it fulgore coruscant

Ardentes auro et paribus lita corpora guttis,

Haec potior suboles; hinc coeli tempore certo

Dulcia mella premes, ... "[38]

Varro refers to these bees as small, variegated and round.[39] Columella says that they are small and slender, have pointed abdomens and are variegated, smooth and mild-tempered,[40] while Pliny describes them as short, variegated and compactly round.[41]

The first three abdominal rings of the Italian bees vary in colour from a bright golden yellow to the brownish yellow of ochre, and today the darker ones are distinguished as "leather-coloured" Italians. It may be that Columella, after describing black or brown bees as "infusci coloris", referred to these leather-coloured bees as "coloris meliusculi". The descriptions of the various kinds of bees other than the yellow-banded ones however is vague and inaccurate. The darker bees were evidently unpopular, and from the fact that the slender bands of hair between the abdominal segments is more prominent on them than on the yellow bees: the former are variously described as of rough or objectionable appearance - e.g. "turpes horrent" (Vergil); "horridi pili" (Columella); "deterrimae ex iis pilosae" (Pliny).

The dark races of bees are of practically the same size as the Italians; - at least no difference is apparent to the eye provided the bees are equally fed. Several references are made, however, to supposed differences in size, and Columella, quoting and supplementing Aristotle refers to: -

" ... alias vastas, sed glomerosos easdem nigras et hirsutes" and

" ... minores sed aeque rotundas" and again

38	Vergil, Georg. IV, 96-100.	
39	Varro, R.R. III, 16-19.	
40	Colum. R.R. IX, 3.	
41	Pliny, H.N. XI, 19.	

" ... alias, magis exiguas, nec tam rotundas", and lastly

" ... nonnullas minimas".[42]

Aristotle speaks of a large kind of bee resembling the hornet, "*ἀνθρήνη*", and of another which he names "*φώρ*", - black and broad in the abdomen. The latter is probably the ordinary bee which has become a robber. It is a common thing to see old robber bees at the openings of hives seeking a chance to enter, their backs being black and broad-looking on account of a complete absence of the pubescence which they once had, but have lost.

The bee "like a hornet" is more difficult to explain. The hornet is a large species of wasp and it is quite likely that it may have been at one time mistaken for a kind of bee, especially as wasps are often found in hives which they have entered for plunder during the late summer and autumn. It is possible therefore that Aristotle made this mistake himself or that he copied from an earlier inaccurate writer. In any case there is certainly no honey-bee like a hornet. Pliny refers to bees like wasps as distinct from the variegated bees, -

"Deteriores longae et quibus similitudo vesparum",[43] but he may have got the idea from Aristotle or from a source available to both writers.

Varro, if we accept Schneider's reading, describes a bee that resembles a wasp, is not social in work, is accustomed to injure by its bite and which the bees keep away from themselves.

"Vespae quae similitudinem habet apis neque socius est operis et nocere solet morsu quam apes a se secernunt."[44]

Schneider's emendation "vespae" for "vespa", however, makes Varro say what is absurd. Gesner and others have "vespa" which makes Varro's description apply to the wasp. This seems a much more likely reading, for the wasp certainly does not help in the work of the hive, has a formidable "bite" and is kept at arm's length by the bees.

---

42   Colum. IX, III.
43   Pliny, H.N. XVIII, 19.
44   Varro, R.R. III, 16, 18.

ORIGIN OF BEES

The ancients were quite unable to determine the sex of bees or to discover the manner of their procreation. Sexual intercourse was never observed amongst them, and this is not to be wondered at, for the coition of queen and drone takes place usually only once in the life of the former and then only in the air and at some distance from the hive. Indeed, very few modern apiarists have observed the act itself although it is not uncommon to see a young newly-fertilised queen returning to the hive from her marital flight showing certain unmistakable signs of her fecundation. Moreover, although the Romans were not altogether unfamiliar with observatory hives, - to be referred to later in this paper - they seem never to have observed the queen laying her eggs in the cells of the combs. And so various explanations were made concerning the mode of reproduction of bees. The most widely believed of these was that they collected their progeny from certain flowers. Vergil, for example, following Aristotle, after stating that they never indulge "in concubitu" or bring forth their young by travail, says that they collect their young from sweet herbs, -

> "Neque concubitu indulgent, nec corpora segnes
>
> In Venerem solvunt aut foetus nixibus edunt;
>
> Verum ipsae e foliis natos et suavibus herbis
>
> Ore legunt … "[45]

Columella, always practical, thinks it not worthwhile to consider whether

> "examina … concubitu subolem procreont," or
>
> "haeredem generis sui floribus oligant".[46]

Varro believed that some bees were produced from bees and others from the putrified body of an ox, -

> "Apes nascuntur partim ex apibus partim ex bubulo corpore putrefacto."[47]

(The latter belief, which was very prevalent, will be discussed in a later chapter).

Pliny is candidly puzzled about the origin of bees. He says that they were never seen to mate, -

---

[45] Vergil, Georg. IV, 198-201.
[46] Colum. R.R. IX, 2 .4.
[47] Varro, R.R. III, 16 .4.

"Apium enim coitus visus est nunquam".[48]

And that some considered that they were made from flowers aptly and usefully formed, -

"floribus compositis apte ac utiliter".[49]

But he goes on to say that others believed that one bee, the King and the only male, begat all the others, these accompanying him, not as their ruler, but as females following the male; this theory being supported by the observed fact that without him there was no brood, -

"ideo fetum sine eo non edi."

But, asks Pliny, how can it be that the same union gives rise to the two kinds of bees, - workers and drones?

"Quae enim ratu ut idem coitus alios perfectos, imperfectos generet alios?"[50]

He thinks the theory of the origin in flowers might be the better one but for the difficulty arising from the presence of the drones, - bred in the outer combs.[51] In the end, he avoids expressing a definite opinion on the matter.

There can be little doubt that the pollen was the "fetus" or progeny supposed to be collected from the flowers. Its granular character may have suggested that it was the origin of the numerous "vermiculi" or grubs found in the comb whose numbers evidently increased or decreased in approximate proportion to the amounts of pollen gathered at different times of the year.

Pliny's difficulty respecting the production of two kinds of bees from the same marital union was justified and accounted for when Dzierzon in 1852 discovered the principle of parthenogenesis in bees. We now know that the queen, in the act of laying an egg, may cause it to be impregnated by some of the spermatozoa she derived from the drone with which she mated, in which case the egg produces a female (worker or queen) bee; or she may withhold them from the egg, which produces a male (drone) bee.

The opinion that the young bees were gathered from flowers prevailed throughout the Middle Ages and Royds quotes Rusden, the bee-master of Charles II, who expresses it as late as 1679. The sex of the queen bee and the fact that she lays the eggs from which the other bees are hatched were discovered by the celebrated Dutch naturalist Swammerdam in the latter part of the 17th century.

48	Pliny, H.N. XI, 16.
49	Pliny, H.N. XI, 16.
50	Pliny, H.N. XI, 16.
51	Pliny, H.N. XI, 16.

BROOD

The eggs deposited in the cells by the queen hatch in four days. From each there comes a tiny white larva which the worker bees immediately provide with pre-digested food elaborated by themselves from pollen and honey. The larva literally floats in this at first and grows rapidly, until, on the ninth day after the laying of the egg, it is sealed up in its cell by the worker bees to pass through the chrysalid stage. On the twenty-first day it bites through the capping of its cell and emerges a perfect adult bee. Drones have a longer period in the chrysalid stage and emerge on the twenty-fourth day. When it is necessary for the bees to rear a new queen, they build a few large drooping cells, known as "queen-cells", in each of which the reigning queen deposits a duly impregnated egg. The resulting larvae, which are identical at first with those that produce worker bees, are fed much more liberally with digested food, in consequence of which they develop to a large size, and, after a week in the chrysalid stage, finally emerge on the sixteenth day as fully developed female bees.

Pliny gives us some details about the brood. He says the bees sit, as do fowls (which, of course, is not strictly true, the eggs being hatched by the heat of the hive due to the presence of the bees) and that at first a little white worm, - "vermiculus candidus" is seen lying crosswise and adhering to the waxen cell.[52] As time goes on the bees pour food on to the larvae, -

"Tempore procedente instillant cibos."[53]

The heat necessary for hatching the eggs, he continues, was believed to be generated by the strong humming of the bees, - "maxime murmurantes" [54] and, he adds, the young bees need forty-five days to come to maturity! As has been previously stated, the period between the laying of the egg and the emergence from the cell of the mature worker bee is only twenty-one days. To give weight to his statements Pliny relates that the hatching of the bees had been watched in the hives of a suburban Roman nobleman who had caused these to be made of lantern horn through which the bees could be seen, -

"Spectatum hoc ... alvis cornu lanternae tralucido factis."[55]

In this connection it is interesting to note that Florentinus gives twenty-one days as the period in which bees are generated from the decaying carcass of a dead ox!

52	Pliny, H.N. XI, 16.
53	Pliny, H.N. XI, 16.
54	Pliny, H.N. XI, 16.
55	Pliny, H.N. XI, 16.

"Apes ex bove natae una et vicesima die vivificantur."[56]

BROOD COMB

Showing worker-cells, eggs, larvae and capped brood

"Geometriam figurarum pulchritudinem sine arte, sine regulis, sine circino, nempe figuras sexangulas, et sex laterum et aequalium angulorum apes conficiunt."

Aelian. V, 13

"Certum est … incubant. Id quod exclusum est primo vermiculus videtur candidus, jacens transversa adhaerensque ita ut pars cerae videatur."

Pliny, H.N. XI, 16

---

56   Geop. XV, 2.

The "King" alone was believed not to pass through the larval stage, but, being of the colour of honey and as though made from a choice flower, - "electo flore factus", straightway bore wings, - "statim pinniger",[57] i.e. became a chrysalis or nymph. This opinion was held by the Greeks, according to Columella, -

> "Hyginus quoque, auctoritatem Graccorum sequens, negat ex vermiculo (ut caeteres apes) fieri ducem, sed in circuitu favorum paulo majora quam sint plebeii seminis, inveniri recta foramina quasi sorde rubri coloris, ex qua protinus alatus rex figuretur."[58]

The rest of the multitude of larvae, as soon as they begin to take shape, says Pliny, are known as "nymphs", -

> "Caetera turba cum formam capere coepit nymphae vocantur."[59]

The term "nymph" is still commonly applied to the chrysalid stage of an insect.

The amount of brood in a hive at any one time depends on the fecundity of the queen and the income of honey and pollen. Roughly speaking the brood and pollen occupy the cells of the central combs and the honey is stored in those of the outer ones. But any of the cells may be utilised for breeding or for stores and it often happens that when honey is brought into the hive in large quantities, - as during a "honey flow" - so many of the cells are filled with it that the queen does not find sufficient in which to deposit her eggs, and so breeding is restricted. This circumstance is considered bad by the modern bee-keeper, the obvious remedy being of course, to extract some of the honey so as to increase the capacity of the hive. Palladius describes this condition, but in explaining it he attributes wrong motives to the bees. He says that when there are too many flowers they think only of gathering honey and not of their young, -

> "Solam curam gerendi mellis exercent de prole nil cogitant",[60]

with the result that the whole stock may be extinguished owing to the non-renewal of the brood. For a honey-clogged hive he prescribes a striking but foolish remedy, viz.,

> "Close the entrance of the hive for three days, allowing none of the bees to come out so that they will turn their attention to the production of brood", -

> ". . . ita ad generandum subolem se conferent".[61]

57	Pliny, H.N. XI, 16.
58	Colum. R.R. IX. 11. 5.
59	Pliny, H.N. XI, 16.
60	Pallard. R.R. IV, Tit 15.
61	Pallard. Lib. IV, Tit 15.

Natural History & Economy

Provided the bees were not suffocated, this rather cruel treatment would have the desired effect for some of the stored honey would be consumed without being replenished and the emptied cells made available for the brood. There is, of course, no such thing as "too many flowers" for the wise bee-keeper removes the surplus honey from time to time and enlarges his hives when necessary. To the modern apiarist the idea of imprisoning bees for some days to prevent them from gathering honey is almost grotesque.

Breeding (in temperate climates) continues throughout the year except for two or three months in the depth of winter. During this period the bees were considered to sleep and to take no food, -

> "A bruma (the winter solstice) ad arcturi exortum (Feb. 11th) somno aluntur sine ullo cibo."[62]

In this, however, Pliny is inaccurate, for bees never sleep, although in cold weather they cluster closely together and appear almost torpid, moving only sufficiently to secure the food from the neighbouring cells, - for food is necessary to them at all times.

The length of the lives of bees was not even approximately known to the ancients. Aristotle thinks they live six or seven years and that it is a fortunate colony that lasts for nine or ten years.

Pliny repeats those fallacies,[63] and Columella follows suit,[64] each in turn more positively than the previous writer. Columella states that no swarm can survive more than ten years, for even if new brood be added each year, -

> "Quamvis in demortuarum locum quotannis pullos substituunt."

About the tenth year the population of the whole hive is exterminated, -

> "Nam fere decimo ab internicione anno, gens universa totius alvei consumitur".[65]

Swarms die out from time to time on account of disease, - ("nam saepe morbis intercipiuntur"),[66] starvation, or queenlessness, but not because there is any limit to their period of existence. The writer once purchased some old hives which had been continuously occupied by bees for nearly thirty years. Vergil is more cautious in this matter. The bees live not more than seven years, he says, but the immortal race remains and grandsires of grandsires are counted, -

---

62   Pliny, XI, 16.
63   Pliny, XI, 20.
64   Colum. R.R. IX. 3.
65   Colum. R.R. IX. 3.
66   Colum. R.R. IX. 3.

"At genus immortale manet, multosque per annos stat fortuna domus et avi numerantur avorum".[67]

Columella warns his readers to guard against the depletion of a whole apiary, by encouraging brood-rearing and by catching and hiving the new swarms so as to increase the number of the stocks.[68]

THE QUEEN BEE

If a queen bee be watched through the glass of an observatory hive the following facts will quickly be noticed:-

She is considerably longer than the worker bees; her body is slender and tapering, her wings short in proportion to her body, although actually a little longer than those of the workers; her legs are very long, her thighs erect, and her abdomen more distinguished in markings and colour than those of her progeny.

"Long is her tapering form and fringed with gold.

The glossy black, which stains each scaly fold;

With gold her cuirass gleams and round her thighs

The golden greaves in swelling circles rise".[69]

---

67   Vergil, Georg. IV, 208-9.
68   Colum. R.R. IX. 3.
69   Evans, The Bees I, 404-6.

Natural History & Economy

THE QUEEN BEE

"Duo autem genera eorum; melior rufus quam niger variusque. Omnibus forma semper egregia et duplo quam ceteris major, pinnae breviores, crura recta, ingressus celsior … multim etiam nitore a volgo differunt."

Pliny, H.N. XI, 16.

She moves over the combs in a stately manner, the bees respectfully making way for her, - even by walking backwards if necessary. Half a dozen or more workers are her attendants. They always face towards her and from time to time offer her food from their tongues, - for she has no time to feed herself, especially when engaged in laying as many as two or three thousand eggs in a single day. She does not give the impression of aggressive or commanding royalty, but rather that she is subservient to the will of the workers who ply her with rich food and so stimulate her to lay to the utmost of her capacity for the good of the community. For in the nation of bees, as in that of men, the greater the population the greater the wealth and security.

The bees show the greatest love and respect for their own queen whom they readily distinguish from all others. They appear always to be aware of her presence, and if she is lost or taken away from them the bad news travels in a mysterious way, and within a few minutes to all her children, who thereupon display excessive agitation, rushing here and there, inside and outside the hive, enquiring of each other, gradually losing hope, until in a day or two black despair gives place to a new hope – the hope of being able to rear a new queen from one of the eggs or tiny larvae which should be plentiful in the combs. And so, the modern beekeeper knows the queen bee rather as the mother than the monarch of the bees; as the patient servant of the hive community rather than its tyrannical ruler.

But the ancients had an entirely different conception of the ruler of the hive. Known to them as a "King", he organised and directed the work of the colony and sometimes led his bees in battle array to fight against neighbouring swarms.

Palladius, quoting from Columella, describes the "Kings" as a little larger and longer than the rest of the bees, with more erect thighs but not large wings; of a beautiful and bright colour, light and without hair – unless perchance on the abdomen. The darker and hirsute ones he recommends should be killed and the more beautiful retained, -

"Fusci atque hirsuti, quos oportet extingui et pulchriorem relinqui."[70]

Pliny rightly attributes to the "King" a more stately walk - "ingressus celsior" than that of the other bees, but he is mistaken in saying that he carries on his forehead a bright spot bearing some resemblance to a diadem, -

"in fronte macula quodam diademate candicans".[71]

The prejudice against dark-coloured "Kings" was general. Varro refers to three kinds, black, red and variegated, but quotes Menecrates who recognised only two, the black and the striped, the former of which he recommended to be killed because (as Menecrates supposed) he creates disunion, and either causes the better King to leave with some of the bees or goes off with them himself.[72]

Vergil is extravagant in his condemnation of the dark-coloured "King".

" ... ille horridus alter

Desidia latamque trahens inglorius alvum".[73]

The ancients, like many present-day beekeepers, were doubtless influenced by the superior beauty of the "golden" Italian bees, but their excessive dislike of the darker races was unwarranted by sound experience. The Italian bees are generally considered to be better honey-gatherers than the "blacks" but the latter usually excel in the beauty of their "comb" honey and have hardier constitutions, for which reasons they are preferred by many experienced apiculturalists of today. Moreover, in the United States of America,

---

70 Pallad. VII. Tit. 7.
71 Pliny, H.N. XI. 16.
72 Varro, R.R. III. 16. 18.
73 Varro, R.R. III. 16. 18.

the greatest beekeeping country in the world, where the Italian bees are now the favourite race, the darker "leather-coloured" queens are preferred to the "golden" specimens, as a glance at the advertising columns of any American Bee Journal will show.

All the old writers laud the supposed obedience and undoubted affection the bees have for their "King". According to Vergil, not Egypt itself, nor Great Lydia, nor the Parthians, nor Median Hydaspes are so obsequious to their King. Whilst he is safe there is one mind for all. If he is lost, they break their allegiance and destroy their honeycombs. He is the guardian of their works. Him they admire and humming vigorously stand around him as a guard. Often, they carry him on their shoulders and expose their bodies in war, seeking through wounds a glorious death.[74]

Varro, Didymus and others state that the bees carry their King on their shoulders when he is tired, -

"Fessum sublevant, et si nequit volare succollant."[75]

"Defessum ipsum sustinet, et cum nequit volare, gestat et conservat."[76]

The idea that the bees should thus cherish their King is charming but not true in fact, except that the queen's immediate attendants occasionally stroke her body with their antennae, or remove with their tongues such substances as dirt, pollen, or honey which may accidently have got on to her body, and except also when she is with a clustered swarm; at other times they scrupulously avoid contact with her.

It is possible that bees recognise their own queen by the sense of smell. They examine strangers with their antennae and invariably fall upon and destroy a new queen introduced to them unless they are deceived by one of the various artifices known to modern beekeepers, most of which depend on securing that she first acquire the scent of the hive, or at least that her own characteristic scent be destroyed. The commonest method of introducing a queen to an alien stock is to put her into a wire cage with some food and to place this amongst the bees with whom she is to live. Thus, protected from their attacks she gradually acquires the scent of the hive and may be liberated amongst them after two or three days. The writer has devised a simple method of introducing queens which depends on the destruction of the scent of the queen by immersing her in water for a few seconds. Thus treated, she is readily accepted by a strange and queenless colony. It is probable, however that this is a successful method because it also discourages any hostility the queen may show towards the bees. Aelianus considered that the bees recognise their queen by scent for he says, -

---

74   Vergil, Georg. IV 210-218.

75   Varro, R.R. III, 16. 8.

76   Geop. XV, 3.

"Ipsum autem sagassime odorantur".[77]

The same writer would have it that the "King" has important regal functions and responsibilities:- He cares for the bees and directs their assembly; some he commands to fetch water, - "aquari jubet"; others to mould, to build, to polish and to carry amongst the combs, - "alias inter favos fingere, extruere, expolire, suggere"; and others to set out for the pastures, - "alias ad pastiones proficisci"; the King himself, he says, has quite enough to do to look after these things and to enact the laws, - "leges sancire".[78]

Vergil attributes warlike qualities to the Kings and vividly describes a battle between two swarms headed by their leaders. He pictures discord arising between them; hearts beating fiercely in anticipation of the coming fight; the clarion of brass that summons the loiterers; the voice which is heard like the broken sounds of trumpets; the excitement of the assembly; the glittering of wings; the sharpening of the stings with beaks; the dense crowding around the King at his doors and the loud challenge to the foe; the bursting forth from the gates and the shock of battle; the falling of the slain like hail, or acorns from the shaken oak; the heroic and obstinate efforts of the Kings, distinguished by their wings – who exert mighty thoughts in narrow breasts, each resolved not to yield until the victor shall have put the enemy to flight;- all of which excitement and strife, he adds, is repressed by the throwing of a little dust.[79]

77　　Aelian. V, 10.
78　　Aelian. V, 11.
79　　Vergil, Georg. IV, 67-87.

Natural History & Economy

VERGIL'S BATTLE OF THE BEES

An illustration from Dryden's Vergil, published in 1698. The bees are being calmed by the throwing of dust.

"Hi motus animorum atque haec certamina tanta

Pulveris exigui jactu compressa quiescunt."

Vergil, Georg. IV, 86-7.

Various writers have attempted to explain the battle of Vergil's bees. It is clear that he has utilised some of the phenomena of swarming, -

"Vox auditur fractos sonitus imitata tubarum"

which probably refers to the piping of the young queens, often heard before the issue of a "second" swarm;

"Et circa regem atque ipsa ad praetoria densae miscentur", -

probably an allusion to the clustering of bees outside the hive frequently observed for some days before a swarm leaves a densely populated hive,

"Motus ... atque ... certamina ...

Pulveris exigui jactu compressa quiescunt", -

a reference to the throwing of dust (an alternative to sprinkling with water) to cause a swarm to settle. But bees do not fight when they swarm, - unless indeed they enter a hive already occupied, - which does not agree with Vergil's description. Two swarms issuing from different hives at the same time often mingle and unite without strife. But all these facts do not account for a battle such as the poet describes, and in fact no such phenomenon is ever observed. Nor is it necessary to find an explanation for the details of natural history on the poems of Vergil who could claim a poet's privilege in attributing to animals the emotions and exploits of men.

THE DRONES

In the early summer when the flowers bloom in abundance and the bees begin to bring into the hive more honey than they consume, and when young queens are reared to provide for the "swarming" which results in the multiplication of colonies, the queen mother, guided by the attendant bees, deposits eggs in some of the larger cells of the combs. Being unimpregnated, - as previously explained – these eggs produce the male bees, or drones, as they are called. A few hundreds only are reared as a rule and – from the beekeeper's point of view – the less the better, for their sole use is that of fertilising the young queens, and for this not one in a thousand is ever needed. As the drones do no work, but consume liberal quantities of food, the wise apiarist limits their numbers by arranging that the number of cells (drone cells) in which they can be reared is kept at a minimum. Towards the end of the summer, when honey-gathering diminishes, the drones are not needed and so they are expelled from the hives, deprived of the digested food on which they partly depend and which is supplied to them by the workers and are left to die.

THE DRONE

"Genus amplioris incrementi simillimum apis. Nam neque alimenta congerit, et ab aliis invecta consumit, ... sine industria favis assidens."

Colum. IX, 15, 1.

"Sunt autem magni, aculeoque carentes, et ignavi."

Democritus, Geop. XV, 9.

The drones are handsome fellows, much larger than the queen or the workers. Their bodies are nearly as long as that of the queen but much stouter. They have powerful wings and highly specialised vision to aid them in the competitive chase which results in the fertilisation of a queen by the keenest and fleetest of their number.

They could not take part in the work of the hive even if they would, for their tongues are too short to sip the nectar from the flowers, they have no pollen-collecting apparatus and they cannot secrete wax, -

"Their short proboscis sips

No luscious nectar from the wild thyme's lips,

From the lime's leaf no amber drops they steal,

Nor near their grooveless thighs the foodful meal:

On other's toils in pampered leisure thrive

The lazy fathers of the industrious hive."[80]

Phaedrus in his fable "Apes et Fuci, Vespa judice", makes the drones claim as their own work the honeycombs which the bees had constructed in a lofty oak. The wasp called in as an arbitrator directs both parties to take hives and if they can, construct combs similar to those in dispute. The drones refuse but the bees agree, -

"Fuci recusant; apibus conditio placet"

upon which the issue is no longer in doubt and the combs are duly restored to the bees.[81]

Possibly no creature has been more maligned than the hapless drone. From the remotest times it was known that he possessed no sting and that he took no part in the labours of the hive but lived on the fruits of the others' toil. And so he has always been an object of contempt and satire, and even his name has become synonymous with that of a lazy worthless fellow.

In his "Works and Days," Hesiod (circa 800 B.C.) says, -

"And with him gods and men are indignant who lives a sluggard's life like in temper to stingless drones which lazily consume the labour of bees by devouring it."[82]

It is not only men, however, who suffer by the comparison, as is seen from the poet's amusing denunciation of women, -

"For from her (Athene) cometh the race of woman-kind. Yea, of her is the deadly race and the tribes of women. A great bane are they to dwell among mortal men, no helpmeet for ruinous poverty, but for abundance. And as in roofed hives bees feed the drones which are conversant with the deeds of evil all day long until the going down of the sun, the bees busy them in the day-time and store the white honeycomb, while the drones abide within the roofed hives and gather the labour of others into their bellies, - even so Zeus, who thunders on high, made woman to be the bane of men, to be conversant with the deeds of evil; and in place of a good thing he gave them a second evil."[83]

---

80 Evans, The Bees. I, 349-54.
81 Phaed. III. Fab. 13.
82 Hesiod, Works & Days, (Mair) p. 360.
83 Heo, Theog. P. 590

(circa 400 B.C.) makes Critobulus say that weeds are like drone bees in a hive, which are of no value in themselves and yet live upon the industry of the working bees and destroy the provisions laid up to be manufactured into wax and honey, - to which Socrates replies that weeds should be plucked up, as the drones in a hive are killed and discharged from it.[84].

Aristotle knew that the drone is incapable of work, that it possesses no sting[85] and generally remains in the hive; that when flying it makes much noise, and on returning to the hive gorges itself with honey.[86] He even quotes the view of some people that the drones are male bees,[87] an opinion not confirmed as a fact until more the 2,000 years later.

All the Latin writers on bees refer to the idleness of the drones and to their expulsion from the hives. Varro, for example, states that they do not help in the work, but eat up the honey, and that many of them, crying out in terror, - "vocificantes" - are at times pursued by a few of the bees.[88] Pliny observes that they are expelled from the hives when the honey has begun to mature, - "cum mella coeperunt maturescere",[89] which is quite true. He is in error, however, when he says that they are as though in the service of true bees, - "quasi servitia verarum apium", - who command them, compel them to be the first to perform work and mercilessly punish those who delay.[90]

The drones help also, continues Pliny in promoting the warmth of the hive, - " . . . adjuvant eas (apes) multum ad calorem conferente turba,"[91] an idea which Columella carries still further, for he says that they assist in brood-rearing by sitting on the "seeds" from which the bees are formed, -

> "Ad precreationem subolis conferre aliquid hi fuci videntur insidentes seminibus quibus apes figurantur."

It is commonly recognised that drones contribute to the maintenance of the hive temperature, but the wise apiarist prefers that this be done by a strong force of worker-bees. There is no evidence that the drones take any part in brood-rearing or any other duties of the hive.

84	Xenophon, Oecon, XVII
85	Arist. H.A. V, 22, 1.
86	Arist. H.A. IX, 40, 5.
87	Arist. H.A. V, 21, 3.
88	Varro, R.R. III, 16, 8.
89	Pliny, H.N. XI, 11
90	Pliny, H.N. XI, 11
91	Pliny, H.N. XI, 11.

Varro calls the drone a thief, - "Fur, qui vecatur ab aliis fucus,"[92] an idea which is developed with amusing extravagance by Aelianus, who says that the Drone, born amongst the bees, remains hidden amongst the honey cells by day, but that at night, when it has observed that the bees are asleep, it invades their works and plunders the hives. This it repeats when the bees are foraging by day. When they return, however, they do not spare the robber but fall upon him and angrily smite with their stings until he pays with his life for his voracity and gluttony, of which things, says Aelianus, naively, beekeepers tell and persuade him, - "μελισσουργοί λέγουσι ταῦτα καί ἐμέ πείθουσιν",[93] - a remark which tempts one to conclude that the said beekeepers may have been merely presuming on his credulity.

As drones are much larger than worker bees it is possible to prevent them from returning to the hive, once they have emerged from it, by intercepting them in traps fitted with sheets of perforated wood or zinc, the perforations being large enough to permit the passage of the workers but too small to allow the drones to pass. Although 'drone traps' are a comparatively modern invention, it is interesting to notice that Aristotle states that some people excluded the drones from the hives by means of suitably perforated material.[94] It is significant that none of the Latin writers refer to this practice, possibly because the Roman beekeepers realised that it was both futile and cruel. Today drone-traps are discredited, except by novices, for it is easy to limit the rearing of drones by providing the bees with artificial waxen foundations for their combs, these being impressed with the beginnings of cells of such a size as to be convenient only for the rearing of worker-bees.

One can hardly leave the subject of the drones without referring to the absurd but quaint directions for their destruction given by Democritus: -

The beekeeper is enjoined to sprinkle the insides of the lids of the hives with water in the early evening, -

"Vespere incipiente opercula vasorum intus aqua irrorato,"

and about daybreak to open the hives, when he will find the drones adhering to the drops, - "guttis adhaerentes," for, says Democritus, being often gorged with honey they have an insatiable appetite for water and will not leave the moisture on the hive-lids, -

"ob aquae appetentiam insatiabilem ab humore operculorem non discedent,"

and in this way, he continues, it is possible to destroy them all, - "omnes perdere," so that not even one may escape.[95]

92	Varro, R.R. III, 16. 18.
93	Aelian. I, 9.
94	Aristotle, H.N. XXII, 1.
95	Geop. XV, 9.

THE WORKER BEES

While the drones have always been objects of pity or contumely, the worker bees have on the other hand never failed to excite the unstinted admiration of men. Their unselfish devotion to the work of the community, involving as it does the renunciation of the privilege of motherhood, their patient and arduous toil and their apparent foresight and intelligence have often been extolled as worthy of limitation by mankind. Not until quite modern times was their sex understood and in the 19th Century they were frequently referred to as "neuters". Forbiger in his notes on Vergil's fourth Georgic (1852) speaks of them as "mellificantes apes in quibus mondum ulla sexus nota reperta est."[96] But we now know that they are females incapable of mating, able, however, in certain very rare circumstances to lay a few eggs which produce drones. Their duties are numerous. They gather water, honey and pollen for food and propolis for sealing crevices in their hives; they build the combs from the wax which exudes from their bodies; they provide digested food for queen, drones and larvae; they ventilate and scavenge their hives, always keeping them scrupulously clean; at the entrances they place guards who examine all corners and fiercely repel intruders. They sometimes feed one another and receive loads from those returning from the fields. Generally it is the younger bees who are the nurses of the larvae and do most of the work of the hive while the older ones forage in the fields.

---

96   Georg. IV, Adnott. 1. 197.

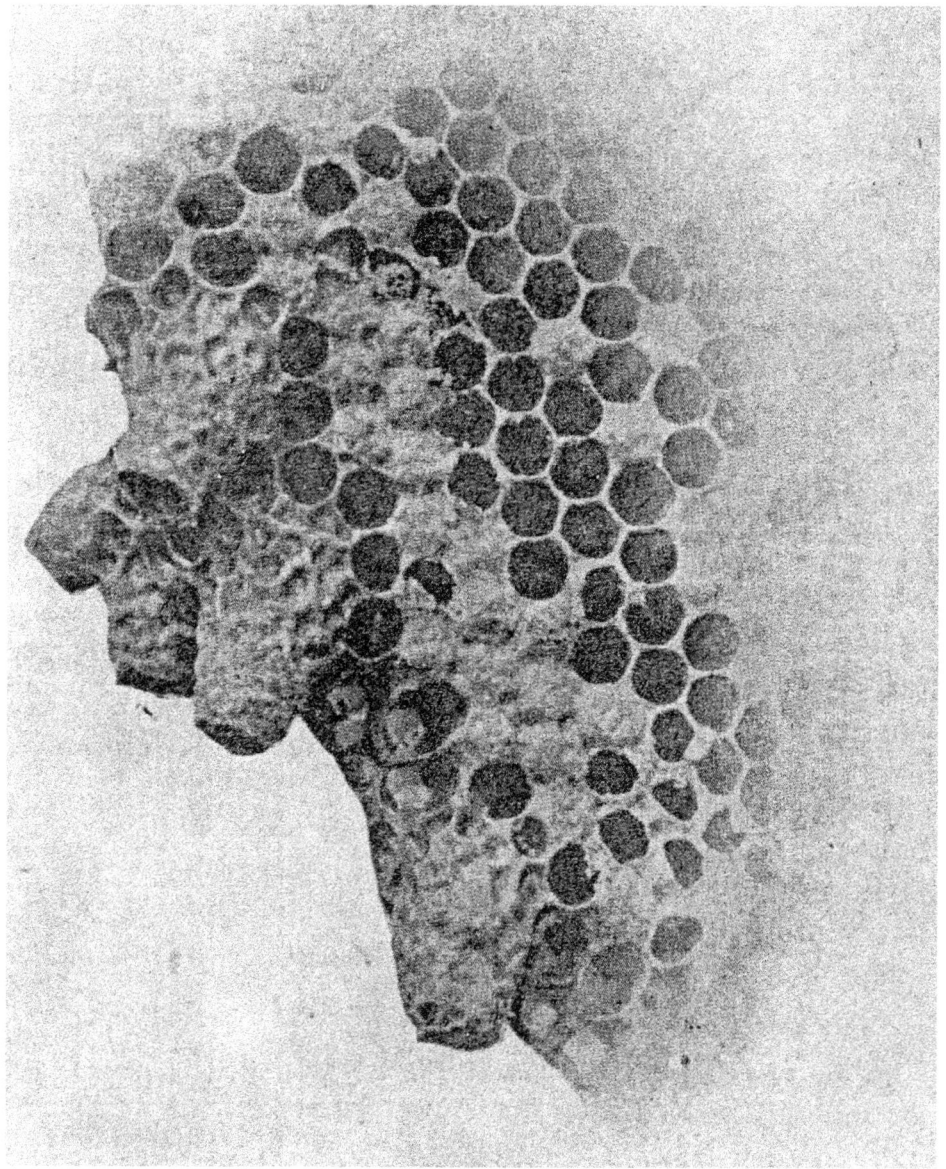

QUEEN CELLS

"Apes (ut aiunt) debemus imitari, quae vagantur et flores ad mel faciendum idoneos carpunt; deinde quidquid attulere disponunt ac per favos digerunt."

Seneca, Epist. LXXXIV.

Natural History & Economy

The ancient writers, from Aristotle onwards described with more or less accuracy what they conceived to be the duties of the worker bees, but by far the best description is that given by Vergil, than which, to the contemplative lover of bees, there can be no more delightful passage in the whole of Bee Literature: -

"Venturaeque hiemis memores aestate laborem

Experiuntur et in medium quaesita reponunt.

Namque aliae victu invigilant et foedere pacto

Exercentur agris; pars intra saepta domorum

Narcissi lacrimam et lentum de cortice gluten

Prima favis ponunt fundamina, deinde tenacis

Suspendunt ceras; aliae spem gentis, adultos

Educunt fetus; aliae purissima mella

Stipant et liquido distendunt nectare cellas.

Sunt, quibus ad portas cecidit custodia sorti

Inque vicem speculantur aquas et nubila coeli;

Aut onera accipiunt venientum, aut agmine facto

Ignavum, fucos, pecus a praesepibus arcent.

Fervet opus, redolentque thymifragrantia mella."[97]

The expression "fervet opus" aptly describes the energy and bustle apparent at the entrance of a beehive on a warm summer day. Great numbers of bees go and return incessantly, scarcely finding room to pass each other in their eagerness to negotiate the crowded entrance, deposit their loads and return to the fields for more. At such a time they are quite oblivious of the presence of the benevolent and fascinated observer and are bent only on the one great object of their lives, that of laying up a sufficient golden store to ensure the continuance of their race through the dreaded winter. Such indeed is "their love of flowers and the glory of making honey" that they often lay down their lives in their work, -

"Saepe ... ultroque animam sub fasce dedere

Tantus amor florum et generaudi gloria mellis."[98]

---

97   Vergil, Georg. IV, 156 – 169.
98   Vergil, Georg. IV, 203-5.

The closest observation of bees has not revealed any organised system under which the various duties of the hive are allocated to particular bees. The ancients liked to believe that there was some discipline and quasi-military organisation, -

> "Protinus autem educti (i.e. the young bees) operantur quadam disciplina cum matribus,"[99]

> "Interdiu statio ad portas more castrorum,"[100]

> "Omnes ut in exercitu vivunt."[101]

The term 'agmen' is frequently used to refer to bees about to work or fight. -

> "Agmine facto ... fucos ... arcent",[102]

> "Cum agmen processit, aliae

> flores adgerunt, etc.,"[103]

and reference has already been made to the organisation of the hive as pictured by Aelianus. To the careful observer, however, there is no sign of authority or direction in the hive. Each bee seems to be a law to itself and to be guided only by instinct to co-operate with its fellows.

Both Varro and Pliny emphasise the supposed hatred which bees have for idleness, the latter going so far as to say that they punish they lazy ones with death, -

> "Cessantium inertiam notant, castigant mox et puniunt morte."[104]

Seneca would have us imitate the unselfish industry of the bees who place what they gather in the combs for the common good, -

> "Apes (ut aiunt) debemus imitari quae vagantur et flores ad mel faciendum idoneos carpunt; deinde quid quid attulere, disponunt ac per favos digerunt."[105]

Several of the ancient writers eulogise the spirit of co-operation which is characteristic of the bees. According to Varro they are a "societas operis et aedificiorum,"[106] they sleep in turns and work alike, -

---

99	Pliny, H.N. XI, 16
100	Pliny, H.N. XI, 16
101	Varro, R.R. III, 16. 9.
102	Vergil, Georg. IV, 167.
103	Pliny, H.N. XI, 10.
104	Pliny, H.N. XI, 10.
105	Seneca, Epist. LXXXIV.
106	Varro, R.R. III. 16 .4.

"alternis dormiunt, et opus faciunt pariter."[107]

Pliny thinks their food and work are equally shared, -

"Neque enim separatim vescuntur, ne inequalitas operis et cibi fiat,"[108]

and that when their day's work is done the humming grows less and less until a single bee flies around and with the same humming with which she awakened the colony in the morning, gives the signal for sleep, as in the manner of a camp, upon which they all become silent for the night[109] for according to Vergil they all have one labour and the same rest from their toils, -

"Omnibus una quies operum, labor omnibus unus."[110]

The views of the ancient writers respecting the government duties, and social relations of bees persisted throughout the Middle Ages and are admirably expressed by Shakespeare in "Henry V,"-

"For so work the honey-bees,

Creatures that by a rule in nature teach

The act of order to a peopled kingdom:

They have a King and officers of sorts;

Where some, like magistrates, correct at home,

Others like merchants, venture trade abroad,

Others, like soldiers, armed in their stings,

Make boot upon the summer's velvet buds;

Which pillage they with merry march bring home

To the tent-royal of their emperor,

Who, busied in his majesty, surveys

The singing masons building roofs of gold,

The civil citizens kneading up the honey,

The poor mechanic porters crowding in

---

107   Varro, R.R. III. 16 .4.
108   Pliny, H.N. XI, 10.
109   Pliny, H.N. XI, 10.
110   Vergil, Georg. IV, 184.

Their heavy burdens at his narrow gate,

The sad-eyed justice, with his surly hum,

Delivering o'er to executors pale

The lazy yawning drone."[111]

COMB BUILDING

Bees construct their combs with wax which is secreted by special glands situated on the underside of their abdomens. Each comb consists of two sets of horizontal, hexagonal cells with a common base between them. A cell in which a worker bee is to be reared is one-fifth of an inch in diameter and about half an inch in depth, so that the total thickness of a comb comprising the two sets of cells and their common base is about an inch. The cells in which drones are reared are usually built towards the outer edge of the combs and are deeper and wider than the worker cells, their diameter being a quarter of an inch. Both kinds of cells are utilised for the storing of honey and pollen when not needed for brood. The queen cells, of which only a few are normally constructed in a season, are much larger than the others. They are usually built at the bottom of the combs, from which they hang downwards. They have thick walls and are an inch or more in length and about one third of an inch in diameter. A square inch of comb contains about fifty-six cells of worker size, or twenty-eight on each side of the common base, and the total number of cells in the combs of a strong stock of bees may be as many as one hundred thousand. The combs hang vertically from the roof of the hive to within about half an inch off the floor, and in places they may be attached to the walls for additional support. There is a space about half an inch wide between adjacent combs which provides a passageway for the bees.

When a swarm is placed in an empty hive the bees suspend themselves from the roof in thickly clustered festoons, and having attained the requisite high temperature through the consumption of food they have carried in their honey-sacs, they are able to secrete wax which they remove in thin scales from their bodies. The process of comb-building is thus described by Cheshire:-

"The wax having been secreted, a single bee starts the first comb, by attaching to the roof little masses of the plastic material, into which her scales are converted, by prolonged chewing with secretion; others follow her example and the processes of scooping and thinning commence, the parts removed being always added to the edges of the work, so that in the darkness and between the bees, grows downwards

---

111   Henry V, 1. 2.

that wonderful combination of lightness and strength, grace and utility which has so long provoked the wonder and awakened the speculation of the philosopher, the naturalist and the mathematician."[112]

Columella describes the combs as hanging from the roofs and adhering a little to the sides of the hive, but not reaching the floor, so as to offer a gangway to the bees. He observes also that the shape of the hive determines that of the combs, -

"Figura cerarum talis est qualis et habitus domicilii."[113]

Pliny gives a similar description but in addition points out that the combs likely to fall are supported by arched supporting walls built up from the floor, -

"Ruentes ceras fulciunt intergerivis a solo fornicatis."[114]

Such short connecting combs, especially when found between the storeys of a modern hive, are commonly known as "brace-combs".

Following Aristotle, Pliny states the order in which the three kinds of cells were believed to be constructed, -

"Domos primum plebei aedificant, deinde regibus. Si speratur largior proventus adiciuntur contubernia et fucis; hae cellarum minumae sed ipsi majores apibus."[115]

It is true that the bees of a swarm first construct the worker cells which form the greater part of the combs; but drone cells are next made and not, as Pliny would have it, the queen cells. Moreover, the drone cells are not smaller than those of the workers, but larger.

The bees store their honey principally in the upper and rearmost parts of their hives. The old bee-keepers knew this and always removed honey from the backs of the hives. As the brood usually occupies the central combs, those in the front are often the last to be filled. Pliny exaggerates this fact and gives a curious but unfounded reason for it. He says that the first three combs are built empty so that the stores may not invite thieves, -

"Prima fere tres versus inanes struuntur ne promptum sit quod invitet furantem."[116]

The regular hexagonal shape of the cells provoked the admiration of the early writers. Aelianus, for example, says that the bees attain the geometrical beauty of the figures, with six sides and six equal angles, without art, without rules and without compasses, -

---

112	Cheshire, Bees and Beekeeping I. 161.
113	Colum. R.R. IX, 15. 7 & 8.
114	Pliny, H.N. XI, 10.
115	Pliny, H.N. XI, 11.
116	Pliny, H.N. XI, 10.

"Sine arte, sine regulis, sine circino."[117]

Varro and Pliny associate the six sides and angles with the number of legs possessed by the bee, -

" ... cum singula cava sena latera habeant quot singulos pedes dedit natura."[118]

Varro tries to account for the shape of the cells by saying that geometricians show that a hexagon is inscribed in a circle so as to include the greatest area, -

"Quod geometrae ἑξάγωνον fieri in orbe rotundo oſtendunt, ut plurimum loci includatur", by which he probably intended to imply that the bees adopt this shape to economise wax in building their cells.

Girard has pointed out that the only rectilinear figures which by continual repetition can completely cover any area are the equilateral triangle, the square and the regular hexagon; and that the last-named figure has a smaller perimeter than either of the others of equal area.

The hexagonal shape, too, is more appropriate than the others for the accommodation of the round-bodied larvae.[119]

We are not justified, however, in attributing the hexagonal shape to the wisdom of the bees. It is probable that at some early stage in evolution the cells were round, as are those of the humble bees today. Round cells in juxtaposition and under the pressure of the contained honey would tend to coalesce and assume the hexagonal shape, and so an instinct for constructing hexagonal cells would be gradually acquired by the bees,

Pliny describes the queen cells as palaces for the future rulers, which the bees build in one part of the hive, - roomy, magnificent, separate and standing out like swellings ("tuberculo eminentes").[120]

117	Aelian. V. 13.
118	Varro, R.R. III, 16, 24.
119	Girard, Les Abeilles, p.72.
120	Pliny, H.N. XI, 12.

A LARGE SWARM

"-… in qua regii generis proles animatur. Est autem facilis conspectu, quoniam fere in ipso fine cerarum velut papilla uberis apparet eminentior, et laxioris fistulae quam sint reliqua foramina quibus popularis notae pulli detinentur."

Columella. R.R. IX, 11, 4.

(Illustration from Root's "ABC" of Beekeeping.)

Columella remarks that a queen cell is easy to see because it is more conspicuous, - "eminentior" and of wider aperture, - "laxioris fistulae" - than the other cells, and because it is situated almost on the outside of the combs. His description of one of these cells - "velut papilla uberis" - is very apt.[121]

Palladius rightly regards the building of the queen cells as the sign of a future "king":

"Est autem hoc futuri regis signum: inter caetera foramina quae pullos continent unum magis ac longius velut uber apparet."[122]

Pliny observes that if a queen cell be squeezed the young queen does not hatch, - "quod si exprimantur non gignuntur suboles,"[123] which may be a reference to the practice, common amongst bee-keepers, of breaking down newly made queen-cells for the purpose of checking swarming.

CLEANLINESS

The hive bee is certainly the most scrupulous of all domesticated creatures in respect of the sanitary conditions of its home. Except in cases of disease, it defecates when flying and when its life is over it dies in the fields. Should it by chance die or be killed in the hive, the other bees immediately remove its body and one of them flies away with it to a considerable distance. Foreign bodies of any kind are not tolerated in the hive and the bees make unceasing efforts to remove them. Unlike the wasps and numerous other insects, they are not carnivorous and do not partake of food that is contaminated.

The earliest observers could not fail to notice these facts and many of the old writers extol the cleanliness and chastity of the bees, at the same time stating, with much exaggeration, that they do not tolerate any foulness or unchastity in the persons of those who tend them.

Varro, for example, quoting mainly from Aristotle, affirms that they will not settle on any place that is foul-smelling or which is scented with unguents, -

"Nulla harum assidit in loco inquinato aut eo qui male olet unguenta;" that they sting a person so scented, - "itaque his unctus qui accessit pungent"; that they do not feed as the flies and no one sees them on flesh or blood or fat, - "Nemo has videt in carne aut sanguine aut adipe"; but they alight only on what has a sweet taste; - "Modo considunt

---

121	Colum. R.R. IX, 11, 4.
122	Pallad. VII, Tit. 7.
123	Pliny, H.N. XI, 12.

in quo est sapor dulcis."[124]

Didymus describes the bee as "purissimum animal."[125] Florentinus includes wine amongst the things which provoke bees to anger, - "Iis tamen qui vinum et unguentum olent saeviores inruunt";[126] and Columella, with whom Palladius agrees, stipulates that the bee-man, when about to manipulate the hives, should be washed, - "nec nisi lotus"; that he should not be a wine bibber, - "neve temulentus"; and that he should abstain from almost all strongly smelling foods such as salt flesh or fish and their gravy, - "Salsamenta et eorum omnia eliquamina," as well as from the stinking pungencies of garlic, onions and the like, - "Item que foetentibus acrimoniis aliis vel ceparum caeterumque rerum similium."[127]

Both Columella and Florentinus absurdly attribute to the bees the ability to discern evidences of recent "Res venereae" in persons. The former says of the custodian of the bees, - "Custodiendum est curatori ... cum alvos tractare debebit ut pridie castus sit ab rebus veneriis",[128] and the latter, - "Mulieres item invadunt maxime eas quae rei venereae operum dederunt."[129] Pliny alludes to equally absurd notions when he says of the bees, - "Praecipitur . ... furem mulierumque menses odere".[130]

Amongst the odours considered to be inimical to bees and therefore to be avoided when a site for the hives was chosen were those coming from swamps, latrines, dunghills and baths.[131] An odour regarded as specially offensive was that of burning crabs, the ashes of which were used in certain medicaments. For this reason Vergil enjoins the bee-keeper, - "neve rubentes ure foco cancros".[132] As a matter of fact, bees take little or no notice of any of the odours which were supposed to disgust or be harmful to them. They are certainly terrified by smoke and the smell of some coal tar preparations such as carbolic acid. The human breath is offensive to them and provokes them to sting; and this is probably the reason why certain foods were believed to be objectionable to them. The gases arising from a marsh might be inimical to their health but even this may be doubted. It is probable that the ancients concluded that odours obnoxious to man were also distasteful or harmful to the bees; but careful observation shows that they appear to

---

124   Aristotle. H.A. IX, 40. Varro, III, 16. 6.
125   Geop. XV, 3.
126   Geop. XV, 2.
127   Colum. R.R. IX, 14. 3. & Pallad. IV. Tit.15.
128   Colum. R.R. IX, 14. 3.
129   Geop. XV, 2.
130   Pliny, XI, 16.
131   Colum. R.R. IX. 5. 1 & 6.
132   Vergil, Georg. IV. 47-8.

be insensible to most of the odours that give us pleasure or pain.

With reference to the care which bees take to keep their hives clean, Pliny rightly says that they remove all things from their midst and no filth lies amongst their work; but, he adds, in order that those working within need not go far from their labours, their excrements are heaped together in one place and removed from the hive on foggy days or when they have little work to do, -

> "Quin et excrementa operantium intus ne longius recedant, unum congesta in locum, turbidis diebus et operis otio egerunt."[133]

The following amusing extract from Butler's "Feminin Monarchi" (1609) sums up, in the main, what the Roman writers thought of the cleanliness and propriety of bees and the need for similar qualities in the bee-master; and incidentally it indicates that, in respect of this part of the subject, at any rate, little or no progress was made in the knowledge of bees between the second and seventeenth centuries: -

> "If thou wilt have the favour of thy Bees that they sting thee not, thou must avoid such things as offend them: Thou must not be unchaste or uncleanly: For impurity and sluttishnesse (themselves being most chaste and neat) they utterly abhore: Thou must not come among them smelling of sweat, or having a stinking breath, caused either through eating of Leekes, Onions, Garleeke and the like; or by any other means: The noisesomenesse whereof is corrected with a cup of beere and therefore it is not good to come among them before you have drunke: Thou must not be given to surfeiting and drunkennesse: Thou must not come puffing and blowing unto them, neither hastily stir among them, nor violently defend thy selfe when they seem to threaten thee; but softly moving thy hand before thy face, gently putting them by and lastly thou must be no stranger unto them. In a word, thou must be chaste, cleanly, sweet, sober, quiet and familiar: So, will they love thee, and know thee from all other."

BEES AND BAD WEATHER

Bees proceed from their hives to forage in warm fine weather. When it is cold or rainy they remain at home, for in cold weather the flowers do not provide nectar, and rain not only impedes their flight but often beats them to the ground, from which they may fail to rise before death overtakes them. Wind also is harmful to them for they are often injured by being blown against houses and trees. They are therefore endowed with surprising skill in divining what the weather is likely to be and so they have always been regarded

---

133   Pliny, H.N. XI, 10.

by beekeepers as good weather prophets. It is an inspiring sight to watch them trooping back to their hives in thousands several minutes before the break of a summer storm.

Aristotle observes that even on a fine day they will keep to the hive if rain or cold weather impends, and that by this sign the beekeepers know that they anticipate severe weather.[134] Varro, referring to their being sometimes overtaken in the fields by a sudden shower or fall in temperature which they have not foreseen, remarks that it rarely happens that they are thus deceived, - "quod accidit raro ut decipiantur." Vergil thinks that it is the duty of the guards at the entrance of the hive to watch the waters and clouds of the sky, -

"Inque vicem speculantur aquas et nubila coeli,"

and he observes that they do not go far from their hives when rain threatens, nor do they trust the sky on the approach of the east winds. He adds that at these times they go for water keeping in safety near the walls of their city and attempting only short journeys, -

"Nec vero a stabulis pluvia impendente recedunt

Longius, aut credunt coelo adventantibus Euris;

Sed circum tutae sub moenibus urbis aquantur,

Excursusque brevis tentant ..."[135]

Pliny also considers that they anticipate winds and showers and then remain indoors.[136] Although bees are undoubtedly aware in a mysterious way of the imminent approach of rain, it is very doubtful whether they know beforehand of impending winds. They are, however, highly sensitive to changes of temperature and the fact that they do not issue from their hives when it is cold accounts for the assumption of the ancients that they could divine the approach of cold winds. It is not uncommon to see them flying abroad quite freely in strong winds if these are warm. Several ancient writers gravely record the curious belief that in windy weather bees carry little stones for ballast, -

" ... et saepe lapillos

Ut cymbae instabiles fluctu jactante saburram

Tollunt, his sese per inania nubila librant."[137]

134	Aristotle, H.A. IX, 40. 25.
135	Vergil, Georg. IV. 191-4.
136	Pliny, H.N. XI, 10.
137	Vergil, Georg. IV, 194-6.

Aristotle would have it that the bees lay down these stones while they drink.[138] Pliny relates that some people think they bear the stones on their shoulders – "in umeris,"[139] but Aelianus considers that they carry them with their feet, - "pedibus portantes," and that their purpose is to prevent the bees from being blown out of their course by gusts of wind, -

"contra ventorum incursiones ... ne ex itinere deducantur."[140]

Pliny adds two interesting and accurate observations on the flight of bees in windy weather. He says that they take advantage of following winds, - "secundos flatus captant," and that against the wind they fly near the ground, avoiding brambles, -

"Juxta vero terram volant in adverso flatu vepribus evitatis."[141]

The notion that bees carried little stones as ballast has no foundation in fact, but it has puzzled some of the commentators of Vergil. Forbiger does not express an opinion. Connington characterises it as pure fiction. Royds thinks that Aristotle or Vergil may have observed bees carrying dead larvae, pupae, or refuse away from their hives and have mistaken these for small stones. Connington is probably right, but there is one possible explanation of the origin of the idea which appears not to have occurred to any of the commentators. Bees sometimes work on flowers of the order "Asclepiadaceae" which grow wild in southern parts of Europe. The pollinia (pollen-masses) of these flowers are provided with a clipping apparatus by means of which they fix themselves firmly to the feet of a visiting bee so as to be carried by it to neighbouring blossoms for the purpose of cross-fertilisation. These pollinia, which remain attached to the bee's feet for a long time, might easily be mistaken by a casual observer for little stones. Just as in quite modern times the pollinia of orchids adhering to the brows of visiting bees, - to be carried thus to neighbouring orchids - gave rise to the belief that some bees were provided with horns. In any case it was only necessary for an error of this kind to creep into Aristotle's writings to be repeated without verification by subsequent writers.

---

138   Arist. H.A. IX, 40. 21.-6.
139   Pliny, H.N. XI, 10.
140   Aelian. I. II.
141   Pliny, H.N. XI, 10.

SWARMING

The population of a hive of bees increases during the spring until it has become so large that the hive cannot conveniently accommodate all the bees. This condition is reached as a rule when there is the greatest abundance of flowers and therefore when food is most plentiful. The bees then prepare for sending out a colony or swarm which must establish itself in a new home. For this purpose they rear a few young queens – usually less than a dozen – and on the day on which the larvae of these are safely sealed up in their cells, that is, when they are nine days old, and seven days before they emerge as mature insects, the mother-queen leaves the hive accompanied by a host of her children varying in number from ten to thirty thousand or more. After circling in the air in joyous excitement for a minute or two, they usually settle in a large cluster on some convenient object such as a branch of a tree, where it pleases the queen first to alight at a few yards distance from their hive. Here they remain for hours during which time they may be shaken without danger into any convenient-sized receptacle which they are usually glad to adopt as their new home. If they are not thus "taken", they fly off again and take possession of a home of their own choosing, - often a hollow tree or the roof of a house. Cowan describes swarming in the following words, -

"There is no mistaking the issue of a natural swarm; the bees leave the hive like a liquid stream; they appear almost frantic, rushing out pell-mell over each other in such large numbers that the atmosphere seems alive with tens of thousands of bees circling around overhead in a condition of joyful enthusiasm which rarely fails to communicate itself to the on-looking bee-keeper."[142]

---

142   Cowan, Guide Book. P. 18.

SWARMING

This swarm, photographed in the writer's apiary in 1920, weighed nearly 7lbs and therefore contained over 30,000 bees. Within two months it stored 100lbs of honey.

Varro and Florentinus give the dimensions of hives which were sufficiently large to throw off swarms of this size, but as a rule the swarms obtained by the Roman methods of beekeeping must have been much smaller.

"Cum agmen glomeratum in proximo frondentis arbusculae ramo consederit, animadvertito an totum examen in speciem unius uvae dependeat; idque signum erit aut unum regem inesse, aut certe plures bona fide reconciliatos."

Colum. R.R. IX, 9, 8.

Evan's description is picturesque, -

> "Mounts the glad chief! and to the cheated eye
>
> Ten thousand shuttles dart along the sky,
>
> As swift through ether rise the rushing swarms
>
> Gay dancing to the beam their sun-bright forms,
>
> And each thin form, still lingering on the sight,
>
> Trails, as it shoots a line of silver light …
>
> Swift as the falcon's sweep the monarch bends
>
> Her flight abrupt; the following host descends,
>
> Round the fine twig, like clustering grapes, they close,
>
> In thickening wreaths and court a short repose."[143]

On the eighth or ninth day after the issue of the swarm, the young queens left behind are ready to leave their cells. As only one is needed to be the future mother of the hive, the first to emerge ranges the combs in search of her unhatched rivals whom she destroys by stinging them while yet in their cells. In this she is helped by the bees who pull down the cell-walls and remove the bodies of the hapless virgins. If, however the hive is still populous, more than one of the young queens is permitted to hatch and these may lead off additional swarms known as second swarms or "Casts". If this does not take place and several queens succeed in hatching safely, there is a series of mortal combats between them which result in one only of their number being left.

Aristotle knew that the "King" left the hive only with the swarm, (he was not aware of the marital flight), and that if he were lost the swarm either returned to the hive or perished and did not store honey.[144] Varro, with the Roman colonial methods in mind, says that the bees send out colonies, - "colonias mittunt"[145] and that the older ones send out the younger, - "progenium veteres omittere volunt in Coloniam," as did the Sabines on account of the great number of their children.[146] He is wrong on this point, however, for the majority of the bees constituting a swarm are the older ones. He describes some of the signs of imminent swarming, -

---

143   Evans. The Bees. II, 295-300 and 313 to 316.
144   Aristotle, H.A. IX. 40. 6.
145   Varro, R.R. III, 16. 9.
146   Varro, R.R. III, 16. 29.

1. A few days beforehand, especially in the evening, the bees hang clustered together like bunches of grapes in front of the entrance, -

"Multae ante foramen (ut uvae) aliae ox aliis pendent conglobatae."

This phenomenon can usually be observed in small, overcrowded hives such as those which were used in Varro's time. It is occasioned by the uncomfortable temperature caused by overcrowding within, and is therefore most often seen in the evening when all the bees are at home. It is not common in modern large and well-ventilated hives.

2. When about to issue as a swarm they make a vigorous humming, like the sound made by soldiers moving camp, -

"Cum jam evolaturae sunt ... consonant vehementer, proinde ut milites cum castra movent".[147] (Palladius says that the uproar goes on for two or three days before the issue of the swarm, and that to detect it, the beekeeper should place his ear frequently against the hive.)

The bees do not make any special sound before swarming, but Varro and Palladius probably had in mind the loud and continuous humming caused by the bees fanning with their wings at the entrance of an overcrowded hive in order to ventilate it.

Many are the directions given to the bee-man to prevent the loss of his swarms. Columella and Palladius emphasise the need for the guardian of the hives to keep constant watch for them, for, -

"Nisi curatoris obsidio protinus excepti sunt, diffugiunt"[148], and

"Novellae apes ... nisi serventur, effugiunt."[149]

Both writers knew that swarms are not to be expected as a rule after about three o'clock in the afternoon and so they recommend that strict watch be kept until about that time, - "in octavam fere diei horam",[150] and "in octavam vel nonam horam".[151]

Columella represents a swarm as eager for a home of its own and contented if this be at once assigned to it by the beekeeper, but if the latter is absent, he adds, it seeks a distant home as though repelled by a sense of wrong – "ut injuria repulsa".[152]

147	Varro, R.R. III, 16, 30.
148	Colum. R.R. IX, 9, 1.
149	Pallad. VII. Tit. 7.
150	Colum. IX. 9. 3.
151	Pallad. VII. Tit. 7.
152	Colum. R.R. IX, 9. 2.

To intercept swarms Pliny recommended the planting of trees near the hives,[153] which was excellent advice; and Vergil specifies the palm or a large wild olive as suitable, -

"Palmaque vestibulum aut ingens oleaster inumbret:

Ut cum prima novi ducent examina reges …

Obviaque hospitus teneat frondentibus arbos."[154]

From the earliest times it has been a custom, whenever bees are kept, to make a continuous clattering noise when a swarm is in the air, to cause it to settle quickly. The practice is now discredited by most beekeepers who know that the bees settle naturally without such an incentive, but although most people doubt its efficacy, there are still some who, perhaps deceived by the fallacy "post hoc, propter hoc," believe in and practise it. It is possible that the "tinnitus", which was usually made by pounding on brass or earthenware vessels, or by clapping the hands, originated in a very early method, used by the owners of hives, of announcing to their neighbours that the bees were swarming and thus laying claim to them.

---

153   Pliny, H.N. XXI. 41.
154   Vergil, Georg. IV. 20-4.

COW DUNG HIVE. SOUTH OF FRANCE

A portion of the Frontispiece of Gesner's Edition of "Scriptores Rei Rusticae," printed in 1735. The picture gives a good idea of the "tinnitus aeris."

Aristotle, referring to the noises by which swarms are collected into their hives, doubts whether they collect on account of pleasure or pain, and succeeding writers are divided in their opinions of the subject. Varro says bees are called birds of the muses because beekeepers bring them back when dispersed by cymbals and clapping, - "cymbalis et plausibus",[155] but later he adds that the effect of throwing dust over them and sounding brass about them is terrifying, -

"Circumtiniendo aere, perterritas perducit,"[156]

Vergil merely recommends the tinnitus and does not venture to state its effects, -

"Tinnitusque cie et Matris quate cymbala circum"[157]

but Columella considers that fugitive bees are terrified by the sounds made with brazen vessels or potsherds and that in their fear they return to their hive or settle on the nearest branch, -

"terreatur fugiens juventus ... eaque vel pavida cum repetierit alvum maternam ..."[158]

Palladius as usual agrees with Columella and says, -

"Examen strepitu aeris terreatur aut testulae".[159]

Lucan thinks that the sound of beaten brass produces astonishment in the bees, -

" ... tum si sonus increpat aeris

Attonitae posuere fugam, studiumque laboris

Florigeri repetunt, et sparsi mellis amorem"[160]

Aelianus, however, thinks the bees are not less appreciative of music and song than the cricket, and that when in flight they are recalled by the beekeepers as though by Sirens, the fabled creatures with bodies of birds and faces of women, who, by their sweet songs, lured sailors onto the rocks to their destruction, -

"Vero tanquam sirenibus retrahuntur".[161]

155	Varro, R.R. III. 16. 7.
156	Varro, R.R. III. 16. 30.
157	Vergil, Georg. IV. 64
158	Colum. R.R. IX. 12. 2.
159	Pallad. VII. Tit. 7.
160	Lucan Pharsalia IX. 288-90.
161	Aelian. V, 13.

Didymus, too, considers that the bee is musical and that it is soothed by sweet melodies, -

"Demulcet autem hoc animal et optima modulatio", -

and that the beekeepers cause them to cluster, when they are dispersed, by the pleasant clapping of their hands, -

"mellarii ipsas dispersas, cymbalis pulsatis, aut manuum concinno plausu congregant".[162]

Although it is doubtful whether bees are susceptible to ordinary sounds, and intelligent beekeepers question the value or the "tinnitus" at swarming time, it is at least interesting to review the opinions of the old writers on the subject, especially as one still meets old-fashioned people who believe that their swarms would desert them unless induced to stay by such agreeable sounds as may be produced with a tea-tray and a door-key.

Another way of persuading a swarm to settle was to scatter dust over them. This method is sometimes used by modern beekeepers, but it is less effective than that of sprinkling them with water. The latter probably causes them to think it is raining and so their ardour for excitement is damped. When they have been thus pacified, say Palladius and have suspended themselves on a branch or in some other place, the beekeeper will know that there is only one "King" if they hang in the form of a single "udder", - "unius uberis eductione pendebunt" - but if there are two or more of these "ubera" then he will know that there are as many Kings and that discord reigns amongst them, -

"Discordes sunt, et tot reges esse, quot velut ubera videris, confitentur".[163]

The beekeeper in such a case is advised to rub his hands with the juice of balm, seek out the Kings and destroy all the dark and hairy ones, retaining only the one which is most beautiful in appearance.

There is, of course, only one queen in a first swarm; but several young queens sometimes accompany a "cast" and then may be seen the almost separate small clusters referred to by Palladius. His advice respecting the choice of a "King", however, was unnecessary and bad, for not only would all the queens except one be killed in a day by nature's method, but the survivor would be the strongest and most vigorous of them all: at least, that is the opinion commonly held by modern beekeepers.

The same writer describes the mode of "taking" a swarm. The bees are drawn by the hand or by a ladle into a new receptacle sprinkled with the customary herbs and honey

---

162   Geop, XV, 3.
163   Pallard. R.R. VII. Tit. 7.

and at evening this is placed amongst the other hives: -

> "In novum vas herbis consuetis et melle conspersum manu attrahatur aut trulla et cum in eo leco requieverit, vespere inter alia collocetur".[164]

The practice of rubbing the inside of a hive with herbs before it is used for receiving a swarm is still followed by some old-fashioned beekeepers who depend on traditional methods rather than those described in modern books. Many of these old methods can be traced directly to Vergil, whose fourth Georgic was regarded as an authoritative work on beekeeping during mediaeval times. Vergil recommends the bee-master to sprinkle the places where it is desired that the bees should settle with bruised balm and the wax-flower plant, the "jussos sapores"; they will themselves settle down in the medicated seats", -

> " ... Huc tu jussos adsperge sapores,
>
> Trita mellisphylla et cerinthae ignobile gramen, ...
>
> Ipsae consident medicatis sedibus".[165]

Florentinus says that the insides of hives should be smeared with the flowers of thyme and white poplar; and that the bees may be persuaded to remain in their hives, these should be daubed with pounded spikenard and myrrh mixed with four times as much honey, -

> "nardi herbae et myrrhae portiones tere, quadruplo mellis admixto, indeque vasa obline".[166]

We do not now think that bees welcome any of the herbs with which their hives were treated. They would certainly be glad to sip the honey prescribed in Florentinus' latter recipe, but on a very warm day this additional food would be quite as likely to send them out again as a swarm as to encourage them to remain; for swarming bees are always gorged with honey sufficient to last them for several days in their new home.

The evil of excessive swarming was recognised by Didymus who warns his readers against taking more than two swarms from one hive in a season. Presumably additional ones were to be returned to the parent hive as is commonly done to-day, -

> "Examina autem ex una vaso plura quam duo non sunt tollenda. Macilenta et debilia erunt".[167]

164	Pallard. R.R. VII. Tit.
165	Vergil, Georg. IV 62-3. 66.
166	Geop. XV. 2.
167	Geop. XV. 3.

ROBBING

When, at the end of the summer, the natural sources of honey diminish, bees tend to depart from their high standard of morality and search for illicit sweets. Then woe betide a weak and disheartened colony whose hive entrance may be large and inadequately guarded. It soon becomes the prey of other bees who enter in ever increasing numbers notwithstanding the fierce opposition of the few guards. In a few days they carry off every vestige of honey, leaving the inhabitants which have not been killed to die of starvation. Once the bees of one colony have begun to rob another, the beekeeper has great difficulty in checking them and the careless dropping of a little honey on the ground near the hives at harvesting time may be sufficient to start robbing throughout an apiary.

Varro says that care must be taken that the strong colonies do not oppress the weak and he gives the excellent advice that a weak colony should be removed from the stronger ones and given a new "King". But his method of dealing with fighting bees, - and it is only when robbing that bees fight, - is the worst conceivable, for he recommends that they be sprinkled with water mixed with honey, -

"Quae crebrius inter se pugnabunt aspergi eas oportet aqua mulsa".[168]

Although this treatment would cause the bees to stop fighting for a time, - that is while they were licking one another, - the honey would attract new robbers to the scene and thus aggravate the evil it was intended to remedy.

Varro adds that if mulsum (which contained wine) were used, the treatment was more effective because the bees applied themselves the more greedily to drink on account of the odour and were stupefied in doing so, -

"propter oderem avidius applicant se, atque obstupescunt potantes".[169]

Pliny observes that when particular stocks are short of honey they invade the hives of others, which in self-defence put themselves in battle-array. He fallaciously adds that, if the beekeeper be present, the side which feels assured of his favour does not attack him, -

"si custos adsit alteruta pars quae sibi favere sensit non adpetit eum.[170]

To stop the fighting he prescribes the throwing of dust, or the application of smoke, milk, or honied water,[171] which remedies, however, would be quite ineffective.

168	Varro, III. 16. 35.
169	Varro, R.R. III, 16, 35.
170	Pliny, H.N. XI, 17.
171	Pliny, H.N. XI, 17.

ENEMIES OF BEES

"Molti nemici ha l'ape, ma il peggiore

E il saputo o ignorante apicoltore",[172] -

"Many enemies has the bee, but the worst is the knowing or ignorant beekeeper", - says Cadolini and it is a fact that "management" by men causes the loss of more bees than all the pests (diseases excepted) to which bees are subject.

Aristotle, Vergil, Pliny and Columella give lists of creatures which are more or less inimical to bees. Of those which attacked them from outside, the hornets and wasps were the most feared, for towards the autumn, being desperate for food, these insects attack and rob badly-guarded hives with impunity, and eat not only honey but the juicy portions of the bodies of the bees as well, -

> "Inter Caniculae et Arcturi exortum (July 26th to September 12th) cavendum erit, ne apes intercipiantur violentia crabronum, qui ante alvearia plerumque obsidiantur prodeuntibus".[173]

Aristotle describes how the wasps were baited in pans containing meat and burnt.[174] Palladius says the hornets become a nuisance in August and that they should be followed up and killed, - but he does not say how. The present-day methods of dealing with wasps viz., - by destroying their queens in the early spring, and by entrapping them with sweet liquor in narrow-necked bottles, appear not to have been used.

Of birds which catch bees in flight and carry them home to their young, are mentioned the merops or bee-eater and the swallow, -

> "Absint ... meropesque aliaeque volucres
>
> Et manibus Procne pectus signata cruentis;
>
> Omnia nam late vastant ipsasque volantis
>
> Ore ferunt dulcem nidis inmitibus escam".[175]

Aelianus recommends that the little birds may easily be captured by making them drunk with meal mixed with wine, -

> "gustatu farinae vino dilutae primum ebriae factae, etc.,"

---

172   Cadolini, cited by B-Bossi &Sartori.
173   Colum. R.R. IX, 14. 10.
174   Arist. H.A. IX, 40. 24.
175   Vergil, Georg, IV. 13. 17.

and he adds that beekeepers moved by the swallow's song do not kill it but are content with preventing it from nesting near the hives.[176]

Several birds not mentioned by the Roman writers are much more inimical to bees than the swallows, amongst them being the great tit, the blue tit, the red-backed shrike and even poultry.

Of harmful animals Pliny names the sheep from whose wool the bees cannot extricate themselves; and frogs, which ensnare bees going to water to drink and are reputed not to feel their stings.[177] Frogs are not now regarded as serious enemies of bees, but toads, which may be included in the term "ranae" used by Pliny, frequently snap up bees, which, returning home tired fall to the ground just before reaching their hives. Lizards were regarded as specially harmful, -

> "Absint et picti squalentia terga lacerti".[178]

Columella recommends the provisions of two or three exits to a hive in order that the bees may escape by one while the lizard is waiting for them at another,[179] – an entirely useless piece of advice for the bees would not recognise an enemy in the reptile nor would they try to avoid it.

Florentinus and Aelianus include land-crocodiles amongst bee-pests, and the latter says that people advised that meal mixed with hellibore should be placed in front of the hives to poison these creatures, -

> "Instruunt farina cum veratro subacta ... ante alveos dispersa ... gustantibus illis perniciem affert."[180]

Quite a variety of insects and other small pests are spoken of, some of which are quite or nearly harmless, such as gnats, -

> "e culicum generi, qui vocantur muliones";[181] the beetles, - "obscaenum scarabei genus,";[182] cockroaches, - "Lucifugae blattae," which devastate the combs on account of the hive entrances being too wide;[183] and spiders, "aranei," which destroy

---

176   Vergil, Georg. IV, 13-17.
177   Pliny, H.N. XI, 18.
178   Vergil, Georg. IV, 13.
179   Colum. R.R. IX, 7. 6.
180   Aelian. I. 59.
181   Pliny, H.N. XI, 18.
182   Colum. R.R. IX, 7. 6.
183   Colum. R.R. IX, 7. 6.

whole hives by the weaving of their webs.[184] But the most serious of the insect pests, mentioned by all the apiarian writers except Varro, was that of the wax-moths. Of these two kinds are well-known to modern beekeepers, -

(a) "Galleria cerella," a small moth, which, having gained admittance to a hive deposits eggs in crevices, combs or debris. When these eggs are hatched, the larvae tunnel through the combs, eating honey, pollen and even the brood of the bees and leaving behind them silken films mingled with excrement. Subsequently they spin cocoons, pass through the nymph stage and emerge as perfect insects.

(b) "Achroia grisella," which has a somewhat similar history.[185]

These little moths invade weakly-populated and bee-less hives and if undisturbed they soon reduce the combs to a disagreeable looking mass of cocoons and excrements.

Pliny describes the damage caused by the wax-moth, -

" ... ceras depascitur, et relinquit excrementa ... fila etiam araneosa quacumque incessit, alarum lanugine intexit".

He does not distinguish the stages of the moth's life and considers that the larvae, "teredines quae ceras praecipue adpetunt," are born in wood, - "Nascuntur et in ipso ligno teredines".[186]

Palladius correctly observes that the wax moths, - "papiliones" - appear first in April, that they abound in May, when the mallows bloom and in these months should be killed.

Following Columella, he advises that a tall narrow-necked metal pot "simile miliario",[187] with a flame burning at the bottom be placed before the hives at dusk. Into this the moths fly, attracted by the flame and unable to escape on account of the narrowness of the mouth of the pot, are consumed.[188] Pliny recommends flambeaux to be placed in front of the hives on moonless nights. The moths were believed to fly out and perish in the flames.[189]

Modern bee-men do not attempt to entrap the moths for they know that these are not tolerated by the bees of vigorous and populous stocks; and by carefully avoiding the exposure of unoccupied combs they prevent the moths from getting a good footing in an apiary.

184	Pliny, H.N. XI. 19.
185	Cowan, Guide Book, 165 – 6.
186	Pliny, H.N. XI, 19.
187	Colum. R.R. IX. 14. 9.
188	Pallad. R.R. V. Tit VIII.
189	Pliny, H.N. XXI. 47.

Florentinus makes a very interesting reference to lice, from which, he says, the bees can be freed by fumigations produced by burning twigs of apple and wild fig trees, -

"Pediculos autem ipsarum tolles, ramulis mali et caprifici incensis ac suffumigatis".[190]

It is most probable that he was referring to the blind louse or "Braula coeca", which sometimes infests the queen bee in considerable numbers and occasionally, the worker bees also. It is not observed by many beekeepers and it seems to do little harm beyond that of impeding the queen's movements. It clings tenaciously but can be dislodged and killed if an infested bee is exposed to tobacco-smoke in a small box. It is doubtful whether Florentinus' prescription would be effective, especially if the smoke were applied to a whole stock at once.

Columella gives general directions for the spring cleaning. The hives were to be opened and fumigated between March and May and all filth and cobwebs removed, -

"Apes curandae, apertis alveis, ut omnia purgamenta ... eximantur, et ut araneis ... detractis fumus inmittatur".[191]

The smoke was to be produced by the burning of dried cow-dung, -

"Fumus factus incenso bubulo fimo",

this being considered specially suitable for bees on account of their supposed origin in the decaying carcases of oxen. If the beekeeper mixed the marrow of oxen with the fuel,

"si fimo medullam bubulam misceas",

the smoke was believed to destroy grubs and moths, -

"Vermiculi quoque . . . et item papiliones".[192]

Pliny says that the smoke of the cow-dung stimulated the bees to activity, -

"Apesque ipsas excitat",[193]

and Palladius says it was good for their health, -

"Fumus incensi (et sicci) bubuli stercoris adhibeatur, qui aptus est apium saluti."[194]

---

190   Geop. XV. 2.
191   Colum. R.R. IX. 14. 1.
192   Colum. R.R. IX. 14. 2.
193   Pliny, H.N. XXI. 47.
194   Pallad. R.R. V. Tit. 8.

Columella recommends that fresh water be used to cool the bees after being smoked and to clean the empty parts of the hive; if some parts cannot be cleaned by water, they may be cleansed by the stiff wing feathers of an eagle or some other large bird, -

> "Couvenit … si quid ablui non poterit, pinnis aquilae, vel etiam cujus libet vastae alitis, quae rigorem habent, emundari".[195]

Large wing feathers still form a useful part of the modern beekeeper's equipment.

DISEASES

Bees suffer from a number of diseases, some of which are not yet clearly understood. The most serious of these maladies are: -

(1) Bee Pest or Foul Brood. Of this there are two kinds, each caused by pathogenic bacteria which cause the death and decay of the larvae in the combs and also infest the adult bees.

(2) Dysentery. Which may be either temporary or malignant.

(3) "Isle of Wight" Disease. Which term may be used for the purpose of this paper to denote diseased conditions characterised by paralysis of the adult bees and the progressive dwindling and ultimate extinction of entire colonies. Dysentery is often a concomitant of these diseased conditions and included amongst them is that now known as "Acarine" disease, recently investigated and described by Rennie.

The ancients were familiar with Foul Brood. Aristotle speaks of a wildness in the bees which causes a strong smell in the hive.[196] Columella in describing it says that the Greeks call it "φαγέδαινα", - a cancerous sore - and that the bees die of it. He gives a fallacious explanation of it, however, for he considers that it begins owing to a paucity of bees caused by the death of many of the workers, either through hard work, or in storms, so that parts of the brood cannot properly be covered by the bees left in the hive. These parts, he says, then begin to decay and this evil gradually spreading and the honey being corrupted, the bees themselves die, -

> " … tuncque vacuae cerarum partes computrescant, et vitiis paulatim serpentibus, corrupto melle, ipsae quoque apes intereant."[197]

---

195   Colum. R.R. IX. 14. 7.
196   Aristotle, H.A. IX, 40. 20.
197   Colum. R.R. IX, 13. 11.

His remedy is to join two stocks together so that the bees will cover the combs; or if that is not possible, to cut out the combs not covered by bees before they begin to putrefy. We now know that the disease is highly contagious and that Columella's remedies must have been almost useless. At the best they could only stave off the death of the affected bees.

Pliny speaks of a condition in which the bees do not bring forth their brood and which was called "Blapsigonia", -

"Vocant ... blapsigoniam, si fetum non peragunt".[198]

This was undoubtedly Foul Brood. Varro, in stating the signs by which bees may be recognised as healthy, speaks of the evenness and smoothness of their work, - "Si opus quod faciunt est aequabile ac leve".[199] He probably referred to the appearance of the capped brood which in a healthy colony is continuous and even, but which, in the combs containing Foul Brood, is patchy and characterised by sunken cappings of the cells containing the dead larvae.

Dysentery, preceded by constipation, has always been recognised as likely to occur in occasional stocks of bees during the spring. When it is not a symptom of a malignant disease it passes off quickly. It has commonly been attributed to the consumption of unwholesome food but at times it cannot be due to this cause and frequently it is one of the principal symptoms of the diseased conditions commonly grouped under the term "Isle of Wight" disease. It is then very serious and is accompanied by heavy mortality of the bees.

That constipated and dysenteric conditions were common in Roman apiaries is evident from the writings of the Latin writers on Apiculture, most of whom refer to it. It was generally believed to be caused by the bees working on particular trees or flowers. Thus Varro ascribes it to the flowers of the almond and cornel, -

" ... morbidae propter primores pastus, qui ex floribus nucis Graecae, et cornu fiunt, coeliacas fieri".[200]

Pliny says that the cornel tree is to be avoided, for having tasted its flowers the bees die of flux, -

"flore degustato, alvo cita moriuntur".[201]

198	Pliny, H.N. XI, 19.
199	Varro, R.R. III, 16. 20.
200	Varro, R.R. III, 16. 22.
201	Pliny, H.N. XXI, 42.

Columella considers that diarrhoea, which he calls "profluvium alvi", and which is fatal unless quickly treated, arises from feeding too greedily after the winter fast, on the tithymalus (sea-lettuce or milk-thistle) and the flowers of the elm, -

> "ita his primitivis floribus avide vescuntur post hibernam famem ... tali nocenti cibo ...",[202]

and he adds that in the parts of Italy where these trees are plentiful, bees rarely live long.

Florentinus advises that the beekeeper remove from the district the tithymallus, hellebore, thapsia, wormwood and the wild fig, for, he says, all these destroy bees because they make bad honey from them, -

> "mel vitiosum ex his faciunt".[203]

Varro's cure for dysentery was to feed the bees with urine. This would be essentially nitrogenous, and it is rather curious to note that Imms, the first systematic investigator of "Isle of Wight" disease, of which dysentery is often a prominent symptom, recommended feeding the affected bees with nitrogenous substances.[204]

Other remedies recommended for dysentery were sorb apples beaten up with honey or urine, or pomegranate seeds sprinkled with Aminean wine;[205] raisins pounded and mixed with sharp wine; rosemary boiled with water and honey; honey containing pounded gall-nuts or dried roses; the perfume of Galbanum; and the root of the Amellus boiled with Aminean wine.[206] The liquid remedies were to be administered by placing them in small wooden or cane troughs – "ligneis canalibus" or hollow tiles, - "imbricis", which were placed either partly or wholly within the entrances of the hives. One occasionally meets old-fashioned beekeepers who attempt to feed their bees in the same way.

It seems clear that most of these remedies were calculated to have an astringent effect, but modern bee-keepers would hesitate to administer them, first because they would be distasteful to the bees and therefore unlikely to be consumed, and secondly because they would almost certainly tend to set up the very evil they were intended to cure. Experience shows that great care is necessary to ensure that bees are fed only with good honey or the best grades of pure cane sugar.

Other symptoms of disease are sufficiently accurately described as to leave no doubt that some of the diseased conditions now popularly known in the British Isles as "Isle

---

202	Colum. R.R. IX. 13. 2.
203	Geop. XV. 2.
204	Imms. P. 140.
205	Pliny, H.N. XXI. 42.
206	Colum. R.R. IX. 13.

of Wight Disease" were prevalent in Roman times and caused beekeepers much anxiety and loss.

The most characteristic of these are worth setting out in detail for it is still popularly believed that the "Isle of Wight" group of diseases is a recent development in the Natural History of bees.

Vergil says the beekeeper may know of disease by the following signs: -

(1) The sickly ones change colour, - "aegris alius color". This is alleged by some modern observers, who profess to see a change to a grayish hue.

(2) A horrid leanness deforms the face, - "Horrida voltum deformat macies". This is purely imaginary.

(3) The dead are carried out of the hives, - "Corpora luce carentum exportant tectis".

(4) The bees cluster together about the entrances of their hives, - "Illae pedibus connixae ad limine pendent", a convincing sign of paralysis characteristic of "Isle of Wight disease".

(5) They loiter in the hive, - "Intus clausis cunctantur in aedibus".

Although affected bees continue to labour while they can fly, the work of the hive rapidly diminishes owing to the increasing disablement and mortality of the bees.

(6) The bees become slothful through hunger and numbed with cold, - "Omnes ignavae fame et contracto frigore pigrae".

It has frequently been a source of astonishment to beekeepers that their bees have entirely died out notwithstanding the presence of abundant and wholesome stores. The cold to which Vergil refers would be caused by the diminution of the population and a first result of it would be that some of the brood would become chilled and die.[207] Pliny and Columella describe approximately the same symptoms, the former adding that when the sick bees are brought out into the warmth of the sun in front of the hives, some minister food to others, -

"cum ante fores in tepore solis promotis aliae cibos ministrant", - an accurate reference to the saddening sight of helpless crawling bees which has in recent years become so familar to British beekeepers.

Columella sites Hyginus who following the older authors, gives a curious recipe for restoring to life the bees which have died of pestilence. They were to be kept dry through

---

207   Vergil, Georg. IV. 251-8.

the winter and about the time of the vernal equinox, brought out into the sun about the third hour of the day and covered with ashes of figwood. In about two hours, animated by the vivifying warmth and having resumed their spirits they would slowly crawl into a hive placed ready for them. But, adds Columella, doubtfully, -

"ipse non expertus asseverare non audeo; volentibus tamen licebit experiri."[208]

Varro gives similar directions for revivifying bees that have been surprised by a shower or sudden cold.[209] It is a common practice of modern bee-keepers to revive benumbed bees by gently warming them, but the figwood ashes, it is hardly necessary to say, are not used.

The condition commonly known as "Spring dwindling" is referred to by the same writer, who says that between the vernal equinox and the 11th of May the stocks begin to increase in strength and population, but those with few or diseased bees perish, -

"intereunt quae paucas et aegras apes habent".[210]

Columella seems to have appreciated the infectious character of bee-diseases for he says that he does not find any remedies other than those he has recommended for other animals, except that the affected stocks should be carried some distance away, -

"nisi ut longius alvei transferantur".[211]

The encouragement of swarming, involving the frequent provision of new hives, which was a feature of ancient beekeeping, tended to restrict bee diseases and it is probable that these were not nearly so serious in Roman times as they are at the present day.

208	Colum. R.R. IX, 13. 3 & 4.
209	Varro, R.R. III. 16. 37 & 38.
210	Colum. R.R. IX. 14. 4.
211	Colum. R.R. IX. 13. 1.

4
MANAGEMENT

HIVES

The natural homes of bees are convenient hollows which they find in trees and rocks. Silenus found his swarm in a hollow elm, -

> "Audit in excesa stridorem examinis ulmo …
>
> Adspicit et ceras …
>
> Atque avide trunco condita mella petit".[1]

Vergil speaks of bees being found in hollows underground and in hollow pumice rocks and trees, -

> "Saepe etiam effosis, si vera est fama, latebris
>
> Sub terra fovere larem, penitusque repertae
>
> Pumicibusque cavis exesaeque arboris antro".[2]

Vergil's commentators have differed in their opinions of "effosis latebris". Forbiger explains,

> "in cavis terrae, non quasi ipsae effodiunt, ut servius videtur accipere. In ea se recipiunt vitandi frigoris causa."[3].

Connington remarks, - "'Effosis' is commonly explained of holes formed by nature or by man. I have been told however that there is reason to think that bees make holes for themselves, which is Servius' interpretation."[4]

To the present writer it seems that all miss the point that while implying that the holes are dug out by the bees; Vergil is merely quoting what others had reported ("si vera est fama"), and which he himself doubts. Honey bees never hollow out homes for themselves nor could they if they would, being unprovided with suitable apparatus; but they always

---

1   Ovid, Fasti, III, 747-52.
2   Vergil, Georg. IV, 42-44.
3   Forbiger, Adnott. Georg. IV, 42.
4   Connington, Adnott. Georg. IV, 42.

occupy ready-made hollow places which are plentiful and easy to find in nature. The mistake of Vergil's informants may well have arisen from observation of certain of the wild bees, e.g. Andrena nigroaenea, which burrow into the ground.[5]

As already stated, a swarm of bees is willing and even eager to accept almost any convenient-sized receptacle as its home. People discovered this in very early times and it was only necessary to provide "alvi" or simple hives and to annex swarms in the woods, to become proprietors of bees. Of the crude hives used in Roman times, several types are still in use in various countries. Varro and Columella give us some detailed information respecting several kinds.

Plaited osier twigs, where available, were commonly used in their construction, -

"Opere textorio salicibus connectuntur".[6]

These were often smeared both inside and outside with cow-dung so that the bees should not be frightened away by their roughness.

"Vitiles fimo bubulo oblinumt intus, et extra, ne asperitate absterreantur".[7]

5	Royds, p. 105.
6	Colum. R.R. IX. 6
7	Varro, R.R. III, 16.16.

CORK HIVES. SPAIN

"Vitiles fimo bubulo oblinumt intus, et extra, ne asperitate absterreantur (apes)."

Varro, R.R. 111, 16, 16.

A hive of this kind without the coating of cow-dung is depicted in illustration No. 16.A., after a Roman bas-relief. It is very similar in appearance to our present-day straw skep and is typical of hives still used in many countries. Dennler, writing in 1920, says, -

> "Dans l'est de l'Europe, et en Asie-Mineure presque toutes les ruches sont constituées par des paniers en forme de dôme dont l'extérieur est recouvert d un centimètre d'épaisseur de boue composée de terre et de bouse de vache. Les abeilles y sont très bien et confortablement installées".[8]

In the south of France, "Bournac" hives are extremely common. They are thus described by Richards (1921), –

---

8   Dennler, in L'Apiculteur, Nov. 1920

"These are made of wicker sticks plaited into a sort of conical circular basket with a diameter of about 18 inches at the base and about 2 feet high. They are plastered over with baked cow-dung with an outside removable cover of straw thatching".[9]

One of these hives is shown in illustration No. 12. There can be little doubt that they are identical in structure with those used by the Romans when Transalpine Gaul formed part of the Empire.

Varro, who had served as a general in Spain, and Columella, who was a Spaniard, both considered that the best hives were those made of cork-bark, a material which was plentiful in Spain and which, being comparatively non-conducting, modified the effects of cold in the winter and of heat in the summer, -

"Haud dubitanter utilissimas alvos faciemus ex corticibus, quia nec hieme rigent, nec candent aestate"[10]

Vergil tells us that the cork bark was sewn, - "ipsa corticibus suta cavatis",[11] and that the hives were to have narrow entrances, for, as he somewhat erroneously adds, cold solidifies and heat melts the honey.

It is interesting to note that cork-bark hives are still in common use in Spain. They are so crude that they cannot have been improved upon since they were described by Varro and Columella.

---

9   Richards, in the B.B. Journal, Dec. 15. 1921.
10   Colum. R.R. IX, 6.1.
11   Vergil, Georg. IV, 33

CURIOUS HOLLOW TREE HAVES

"Haud dubitantur utilissimas alvos faciemus ex corticibus, qui nec hieme regent nec candent aestate."

Colum. R.R. IX, 6.

Mesey Thompson (1921), who took the photograph shown in illustration No. 13 in the South of Spain, says of them, -

> "The hives are simply the bark stripped from a neighbouring cork tree, about four feet long by one foot inside diameter, held roughly together by a piece of wire near the top and another near the bottom. The bees were flying in and out along the seam, the edges of which were not trimmed at all. A disc of cork plugged the bottom. The top is covered roughly by the strip of cork lying on the top. The chief honey harvest is about mid-June, when a knife is run round the inside from above and about 60 lbs of honey taken from the whole of them. A second smaller crop is obtained in September, leaving a little for the bees, which fly through the extremely mild winter, frost and snow being practically unknown."[12]

---

12   Mesey Thompson in B.B.J., August 11th 1921.

Assuming that these hives produced an average of 10 lbs of honey each per year, one can imagine what an enormous number of them (if they were similar) must have been owned by Seius, who is quoted by Varro as having let out his hives at an annual rental of 5,000 lbs of honey![13]

Where fennel grew plentifully, hives were made of the woven stalks, which, as Columella observes, are similar in nature to cork bark. Varro informs us that these hives were square and of about three feet in length and one foot in diameter, -

"ex ferulis quadratas longas circiter pedes ternos latas pedem".[14]

Occasionally hives consisted merely of hollow parts of trees, - "ex arbore cava".[15] Palladius gives interesting directions for obtaining these. If a swarm be found in the branch of a hollow tree, he says, the latter is cut above and below with a very sharp saw and covered with a clean cloth, is carried to the apiary and placed amongst the other hives, -

"si vero in cavae arboris ramo fuerit (examen) acutissima serra idem ramus supra infraque decisus et munda veste coopertus poterit afferri et inter alvearia collocari".[16]

An example of this, photographed by the writer in an English cottage garden, is shown in illustration No. 14. Columella speaks of hives made of boards, - "ligno ... in tabulas desectae fabricantur".[17] But these were not like our modern wooden hives for Palladius informs us that they were shaped like barrels or wine pipes, - "ex tabulis, more cuparum".[18]

13	Varro, R.R. III, 16. 10.
14	Varro, R.R. 888. 16. 15.
15	Varro, R.R. III, 16. 15.
16	Pallad. R.R. V. Tit 8.
17	Colum. R.R. IX, 6. 1.
18	Pallad. R.R. XXXVIII.

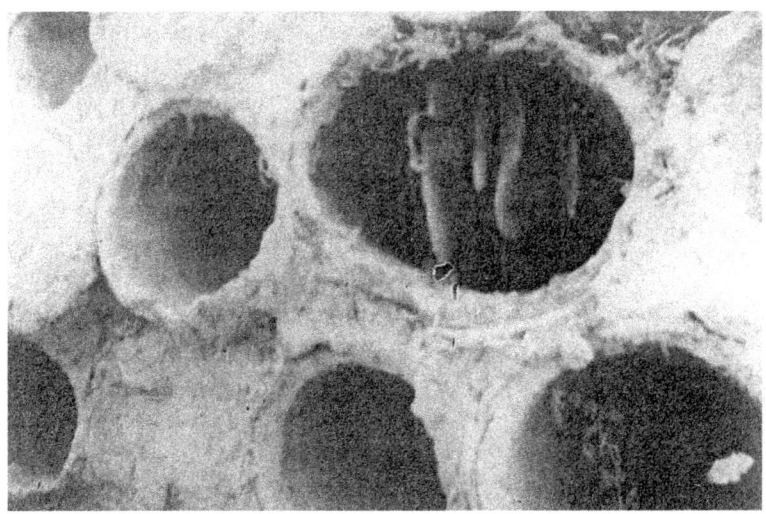

EARTHENWARE AND MUD HIVES. CYPRUS

Photographed by the writer in an English cottage garden. The beekeeper has adopted the plan outlined by Palladius, -

"Si vero in cavae arboris ramo fuerit (examen) acutissima serra idem ramus supra infraque decisus et munda veste coopertus poterit afferri et inter alvearia collocari."

Pallard. R.R. V, Tit. 8

EARTHENWARE AND MUD HIVES, CYPRUS

"Fictilia deterrima sunt quae et hieme gelantur, et aestate fervescunt."

Pallad. R.R. I, 38.

By common consent, however, the worst hives were those made of mud or earthenware, - "Deterrima est conditio fictilium",[19] which were too hot in summer and too cold in winter. Notwithstanding their defects, these were cheap and easy to make and there can be little doubt that they were extensively used in Roman times. Enormous numbers of them are still used in Cyprus and Egypt and examples of both mud and earthenware hives in the former country are depicted in illustration No. 15.

Of the mud hives Carpenter says, -

> "They are made of earth mixed with a kind of chaff and moulded out like a round chimney pot, about 12 inches in diameter at the back, tapering down to the front to about 9 inches in diameter and are 2 feet long. They pile them one above the other, with earth between and they seem to last for two or three seasons. The front lid is made fast with mud and has a tee-hole in the bottom of about 2 inches by ½ inch; the back, sometimes made of wood or slate, is wedged in and the spaces filled with mud. The bees are always put in from the back and usually build their combs at an angle of 45° deg. with the front lid".[20]

Celsus, cited by Columella, disapproves of hives made only of dung ("ex fimo") as liable to take fire, from which it may perhaps be inferred that litter was included with the cow dung in their making. Gesner thinks that if Celsus could live again he would approve still less of the elegantly woven German hives made entirely of straw.[21]

Perhaps the strangest hives in use however were those made of brickwork, - "lateribus extruantur". These were favoured by Celsus notwithstanding their immobility, but Columella rightly condemns them, - "hoc (genus) maxime vitandum est", - for as he points out, they could not be sold, nor could they be erected in other places when the bees were troubled by disease, sterility, or insufficient room.[22]

Reference has already been made to an observatory hive made of horn, spoken of by Pliny. According to the same writer many people made hives of a transparent kind of stone, possibly talc or mica, through which they might observe the bees at their work within, -

> "Multi eas (alvos) et e speculari lapide fecere, ut operantes intus spectarent".[23]

In view of this statement, it is rather remarkable that the general knowledge of the time with regard to the economy of the hive was not more accurate.

19	Colum. IX, 6. 2.
20	Carpenter, B.B.J., Jan. 26, 1922.
21	Gesner, Adnott. Colum. IX, 6, 2.
22	Colum. R.R. IX. 6, 2 & 3.
23	Pliny, H.H. XXI, 47.

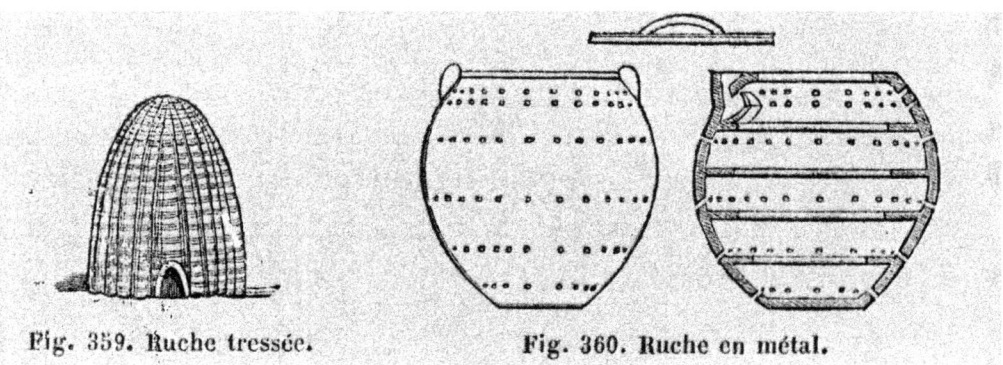

Fig. 359. Ruche tressée.   Fig. 360. Ruche en métal.

ROMAN HIVES

A Woven Hive. "Opere textorio salicibus connectuntur"

Colum. R.R. IX. 6.

Picture to go here No.16.B.

ROMAN HIVES

B. Metal hive with stages and numerous entrances. Found at Pompeii.
From the "Dictionnaire des Antiquités romaines et grecques."

Antony Rich.

The Roman hive shown in illustration No. 16.B is of exceptional interest. It was made of metal and was found at Pompeii. Rich describes it as follows:-

"L'on a découvert a Pompei une ruche artificielle en métal dont on voit … l'extérieur et l intérieur divisé en étages (fori) auquels donnent accès un grand nombre d'ouvertures".[24]

On account of its conductivity, metal is, of course, a most unsuitable material for a bee-hive, and the Pompeian hive was very badly designed on account of the large number of unnecessary entrances. Its chief interest lies in the fact that it embodies what is considered to be quite a modern idea, namely that of working a hive with upper stories or "supers" which are added one at a time as the bees increase and more room is needed for honey storing.

---

24 Rich, Antiquités. P. 304.

Some of the Roman hives were much better than the common straw hives of the present day. Those of long shape were placed horizontally and the lids ("opercula") at the back could be moved inwards until they were near the combs in Winter, or outwards so as to enlarge the hives in Summer, thus fulfilling the purpose of the division-board in a modern frame-hive, -

> "Opercula intus usque ad favos admovenda sunt, omne vacua parte sedis exclusa, quo per hiemem facilius concalescant".[25]

Pliny says that the lid was drawn back little by little as the work was increased, -

> "id paulatim reduci, fallente operis incremento".[26]

The entrances to the hives were made very small to exclude pests and to protect the bees against great cold or heat. Columella goes so far as to say that the entrances, of which there were two or three to an ordinary hive, should be only so wide as to admit one bee at a time, - an arrangement which must have involved cruelty in hot weather.

For protection against the weather in winter, hives were covered with clay, leaves, or straw, and all crevices or cracks were smeared with a mixture of mud and cow-dung. In order that rain should not flow in through the entrances, Columella advises that the hives should be slightly inclined towards the front, -

> "vestibula promiora sint quam terga".

Thoughtful bee-men do not neglect this precaution today for not only is the entry of water prevented thereby, but the sloping floor greatly helps bees which may be engaged in removing débris from the hives.[27]

In contrast with his usual accuracy in respect of practical matters, however, and apparently oblivious of the accepted fact that bees were not carnivorous, Columella seriously records a disagreeable practice followed by some for the conservation of warmth in hives during the winter. He says that killed and disembowelled birds were placed within them so that their feathers might offer warmth to the bees and that should their stores be all consumed they might feed on the carcase, leaving only the bones.[28]

25	Colum. R.R. IX, 14, 7.
26	Pliny, H.N. XXI, 47.
27	Colum. R.R. IX, 7, 4 & IX, 14, 14.
28	Colum. IX. 14. 14.

SITE FOR THE HIVES

Although bees will accept almost any location for their home, the wise apiarist takes care that they are not unduly exposed to cold winds in winter or to the fierce heat of the sun in summer. The ancient writers gave meticulous directions for the choice of the site for the hives.

Columella advised the most secret or retired places, free from noise and the assemblies of men and animals and open to the sunny and least stormy aspect of the sky, -

"Pabulationes ... sint secretissimae, viduae pecudibus, aprico, et minime procelloso coeli statu"[29] and "procul tumultu, et coetu hominum et pecudum."[30]

Vergil would avoid sheep, butting goats and the straying heifer, all of which would not only be likely to overturn the hives, but would injure the pasturage of the bees. The access of winds was to be avoided because they prevented bees from carrying home their loads, -

"Nam pabulum venti ferre domum prohibent".[31]

Columella recommends the bottom of a valley so that when empty the bees may more readily reach their pastures on the higher grounds and when loaded the more easily fly down the slopes towards their hives, -

"Sedes ... sit ima parte vallis, ut vacuae ... facilius editioribus advolent, et collectis ustensilibus, cum onere per proclivia non aegre devolent".[32]

The valley should be near to the house so as to be under the eye of the owner and where it could be reached without difficulty, for, says Columella, beekeeping demands the greatest integrity - "res ista maximam fidem desiderat" - and since this is very rare, the apiary is more safely guarded by the intervention of the owner, - "interventu domini tutius custoditur",[33] from which we may infer that bee-gardens were liable to pillage either by dishonest slaves who tended them or by vagrant thieves.

Varro, too, considered that the hives should be situated near the house and he speaks of some owners, who, for greater security placed them in the portico of the villa itself, -

---

29   Colum. IX. 4. 1.
30   Colum. R.R. IX. 5. 1
31   Vergil, Georg. IV, 9-10.
32   Colum. IX, 5. 1.
33   Colum. R.R. IX. 5. 2.

"in villae porticu",[34] or under the eaves, "subter subgrundas".[35]

Columella says that he remembered sites being given to bees in the hollowed-out walls of the house, - "in ipsis villae parietibus excisis", or in protected galleries, - "in protectis porticibus".[36]

In the porticos the hives were sometimes placed on projections or in niches of the wall, - "in mutilis parietis",[37] and arranged in three rows one above the other. A fourth row was considered undesirable. Spaces were left between adjacent hives so that any one might be manipulated without disturbing the next.[38]

In many parts of the world where apiculture is practised in a crude way, stocks of bees are still kept in holes in the walls of houses. The bees fly outwards and honey is removed from the insides of the walls. In Cyprus, according to Carpenter, even the walls of churches sometimes accommodate swarms. -

"Every few feet of the wall has a stone removed and in these spaces the bees build comb and establish themselves. I was told that the products are sold to support the priest but the bees appear to be public property."[39]

The porticos of Varro correspond, more or less, to the pent-houses in which bee-hives are often kept in modern times.

When situated near the house, the apiary was often surrounded by a wall as a protection against thieves and fire. At a height of three feet from the ground, openings were made in it to serve as flight-holes for the bees, - "tribus elatis ab humo pedibus sit pervius".[40] If there were no need for protection against marauders, the wall was built three feet high and of the same width and the hives placed on the top.

To guard against the depredations of lizards, snakes and other noxious animals the wall was to be made unclimbable by means of smooth plaister or cement, -

"diligenter opere tectorio levigari".[41]

The apiary was to be situated away from bad odours arising from latrines, manure-heaps and baths, -

---

34	Varro, R.R. III, 16, 15.
35	Varro, R.R. III, 3, 3.
36	Colum. IX, pref., 2.
37	Varro, R.R. III, 16, 16.
38	Varro, R.R. III. 16. 16
39	Carpenter, B.B.J. Jan. 26 1922.
40	Colum. R.R. IX, 5. 3.
41	Colum. R.R. IX, 7. 1.

"in ea parte quae tetris latrinae, sterquiliniique, et balnei libera est odoribus".[42]

Perhaps as much for the sake of the bee-master as for that of the bees.

It was doubtless the exception rather than the rule for special structures to be provided to accommodate the hives, for these were commonly kept in fields, orchards and gardens as at the present day.

Most of the ancient writers advise that the hives be turned towards the east or southeast where the sun rises in the spring and winter.

Florentinus gives as the reason, -

"Hibernum aut vernum solis ortum spectare debet, quo et hieme calfiant et aestate aurae inspirantes ipsas recreent".[43]

Columella, however, gives a better reason, viz., that if the hives face the southeast the bees get the warmth of the morning sun when they issue first and are more thoroughly awakened, -

"ut apricum habeant apes matutinum egressum, et sint experrectiores".[44]

Thoughtful present-day beekeepers turn their hives towards the east or southeast so that the bees may be stimulated to early work by the light at the entrances.

Near to an important apiary was situated the "tugurium", a building in which the custodian lived and kept his tools and empty hives.[45]

The ancients rightly laid stress on the need of a good water supply for the bees. Florentinus observes that pure water keeps them healthy and makes good honey, -

"sanas conservat et bonum mel facit",[46]

and Columella says that without it they can fashion neither combs, nor honey, nor brood.[47]

A running stream or some clean still water was an important feature of the apiary, -

"Adsint et tenuis fugiens per gramina rivus".[48]

42	Colum. R.R. IX, 5. 1.
43	Geop. XV, 2.
44	Colum. R.R. IX, 7. 5.
45	Colum. R.R. IX. 5, 3.
46	Geop. XV. 2.
47	Colum. IX. 5. 5.
48	Vergil, Georg. IV, 18.

WATERING-PLACE FOR THE BEES

An illustration from Dryden's Vergil, 1698.

"At liquidi fontes et stagna virentia musco

Adsint et tenuis fugiens per gramina rivus.

…

In medium, seu stabit iners, seu profluet humor,

Transversas salices et grandia coniice saxa

Pontibus ut crebris possint consistere …"

Vergil, Georg. IV, 18-19 & 25-27

Varro recommends that the water be not more than two or three finger-breadths ("duo aut tres digitos") in depth and that into it be placed tiles or small stones, projecting a little above the surface, upon which the bees may settle and drink without danger of being drowned. Vergil thinks they need frequent bridges made with large stones and transverse willows in order that they may rest upon them and spread their wings to dry in the warm sun if perchance the east wind has blown them into the water, -

"Transversas salices et grandia coniice saxa,

Pontibus ut crebris possint consistere et alas

Pandere ad aestivum solem, …"[49]

In the absence of a natural supply, water was to be given by hand, either by conveying it in specially made conduits,

- "extructo canali manu detur"[50]

or by drawing it from a well and placing it in clean wine-press coolers or watering vessels, -

"ex puteo hauriatur in lacus torculares purgatos aut in vasa aquaria".[51]

In the British Isles the natural supplies of water are usually sufficient for bees; but some places are deficient in them and in times of drought the bees may be compelled to fly long distances in search of clean water, or alternatively may be obliged to drink in contaminated places. Good bee-men therefore ensure an adequate supply of pure water by providing it in suitable drinking vessels not far from the hives.

The curious but unfounded belief that echoes were inimical to bees was very prevalent amongst the ancient bee-men. Varro says that the hives should be placed "ubi non resonant imagines", for echoes were believed to cause the bees to take flight. Vergil advises the beekeeper not to trust in a location,

" … ubi concava pulsu

Saxa sonant, vocisque offensa resultat imago"[52]

Pliny says that the effect on the bees is to frighten them, -

---

49 Vergil, Georg. IV, 26-8.
50 Colum. R.R. IX, 5. 5
51 Florent. Geop. XV, 2.
52 Vergil, Georg. IV, 49-50.

"echo ... qui pavidas alterno pulset ictu".[53]

It is probable that this notion arose from the fact the echoes often frightened people and sometimes, perhaps, even put them to flight; and bees were credited with many human attributes and emotions.

Once the site of the apiary was selected, the beekeeper, following the advice of his literary authorities, planted bushes in front of the hives to intercept the swarms, trees to provide shade and flowering plants known to be frequented by the bees. In common with mediaeval writers and even many bee enthusiasts of to-day, the ancients believed in the advantage of growing flowers near the hives, without realising that bees obtain most of their food from flowers at a distance. Pliny believed that they confined their ordinary labours to an area included within sixty paces from their hive, and that if forage failed them there, they sent out scouts (speculatorēs) to search for more distant provisions.[54] We now know, however, that they fly freely up to two miles or more in all directions, and consequently the honey in a single hive may be gathered from an area of twelve or more square miles. Unless, therefore, large crops of particular honey-producing plants can be grown, the cultivation of flowers near an apiary does not appreciably affect the honey-harvest.

As Cheshire says, -

"Cultivating a few melliferous flowers near the hives, after the manner of some amateurs, is scarcely more likely to increase the weight of the supers than growing wheat in a flower-pot is likely to cheapen bread."[55].

STOCKING THE HIVES

Having chosen the site for his apiary and provided himself with hives, the prospective Roman beekeeper concerned himself with the problem of securing his first swarm of bees. He might, indeed, fortuitously become an owner of bees for, as often happens in our own times, a vagrant swarm might settle on one of his trees or take possession of one of his empty hives. Otherwise three courses were open to him: he might obtain a stock from a neighbour, hunt for a bee-tree in the woods, or set a decoy hive in a neighbourhood where swarms were frequent; and for any of these methods he could obtain ample directions from the books of Columella or Palladius. The advice which the former gives a would-be purchaser is excellent; he should open the hive and consider

---

53   Pliny, H.N. XI. 19.

54   Pliny, H.N. IX. 8

55   Bees and Beekeeping. P. 360.

whether it was well-peopled,

> "earum frequentiam, prius quam mercemur, apertis alvearibus consideremus;"

If this were not possible he should see that there were many bees at the entrance, -

> "ut in vestibulo januae complures consistant;"

He should hear a strong sound of humming within the hive, -

> "vehemens sonus intus murmurantium;"

And, if the bees had all retired for the night, he should judge of their multitude or paucity by breathing vigorously into the hive entrance and noting the strength of their angry response, -

> "labris foramini aditus admotis, et inflatu spiritu ex respondente earum subito fremitu, poterimus aestimare vel multitudinem, vel paucitatem".[56]

The modern purchaser of bees would in addition assure himself that they were free from disease, but Columella's tests would usually be sufficient, for diseased hives are seldom vigorous and populous.

The notion that bees should be obtained from the immediate locality rather than from a distance because they are irritated by a new climate, - "solent coeli novitate lacessiri",[57] is quite unfounded. Bees readily settle down in a new position provided this is outside their old range of flight.

Columella's advice respecting the moving of a stock of bees is only partly good. It is best, he says, to shut them in and carry them on one's shoulder by night, -

> "optime noctibus collo portabuntur",

avoiding rough places which may shake them; to rest and feed them during the day while still confined; not to open the hives at once if they arrive during the day but to wait for evening; and to observe for three days whether they show signs of flight, - presumably they were to be confined again if they did.[58]

The injunction to carry the bees gently on the shoulder by night is excellent, but in the case of transportation over long distances, to which alone Columella appears to refer, - "si necesse habuerimus longinquis itineribus advehere", - there would not be the slightest need to confine the bees and to feed them during the day, or to refrain from liberating

---

56   Colum. R.R. IX, 8, 1 & 2.

57   Colum. R.R. IX, 8, 2.

58   Colum. R.R. IX, 8, 3 & 4.

them immediately after their arrival. If, however the bees were moved through a distance less than they had been accustomed to fly, then the plan of confining them for one or more days would lessen their tendency to return to their old home. Palladius practically repeats Columella's directions, adding that if the bees showed a tendency to desert their new location, it was believed that they might be deterred by cow-dung smeared over the hive entrances, -

> "Tamen creduntur non fugere, si stercus primogeniti vituli adlinamus oribus vasculorum".[59]

Swarms of bees often occupy empty hives without the intervention of the beekeepers, and although it is generally considered a dishonourable practice, "decoy" hives are sometimes placed in gardens to attract vagrant swarms. Columella advises his readers to place such hives in the apiaries because some swarms at once seek a home near at hand, -

> "statim sedem sibi quaerant in proximo".[60]

As at the present time the woods often abounded in bees and from Columella and Palladius we gather that it was not uncommon for people to place "decoy" hives near places where bees were observed to visit watering places in large numbers, -

> "Per nemora non longe a fontibus disponunt (alvos) eaque cum repleta sunt examinibus domum referant".[61]

The hives were previously rubbed with herbs, especially balm and honey-wort and sprinkled inside with very dilute honey. But unless a neighbourhood was heavily populated with bees it was not worthwhile to try to catch them in this way, for passers-by often stole the empty hives. Yet, says Columella, the loss of several empty hives is more than made good if one or two swarms are captured.

But a more ingenious and fascinating method of securing the wild bees of the woods is described by Columella and, in less detail, by Palladius. In the month of April and early in the morning the bee-hunter ("indagator") proceeded to sunny places ("apricis locis") in the woods and searched pools or streams where many drinking bees could be found, -

> "Loca mellifica indicant apes, si circa fontes frequentissimae pascantur".[62]

Having provided himself with a small quantity of a red liquid into which he dipped a twig, he used this to mark the backs of some of the drinking bees. After watching their

---

59	Pallad. R.R. I, 39.
60	Colum. R.R. IX, 12, 4.
61	Colum. R.R. IX, 8. 13.
62	Pallad. R.R.V. Tit. 8.

departure and noting their return he judged the distance of their hive from the spot by the length of time they were absent. If they returned quickly, it was only necessary to watch carefully the direction of their flight to discover their home. But if they were long in returning their hive was known to be distant and another method of finding them was used. A piece of hollow cane cut so as to include two nodes and provided with a hole in the side was smeared on the inside with a little thin honey or sweet wine and placed near the drinking bees. The sweet contents would attract them and when a number had entered the hunter closed the hole in the cane with his thumb. In a minute or two the imprisoned bees would be in great distress and only anxious to return to their hive. By removing his thumb a little the hunter allowed a single bee to escape and after intently watching its flight, he followed it as far as he could. He then liberated another and followed it likewise and by repeating the process finally arrived at the dwelling, - usually a hollow tree - occupied by the bees. Once found, they were expelled by smoke and, according to these writers, caused by the aid of the "aeris strepitus" to settle on a brush or a branch of a tree, whence they were received into a prepared hive and carried off to the apiary.[63]

Although the mode of dealing with the bees when discovered was bad, the method of finding them was quite feasible. Indeed, it is still practised in parts of the American Continent where "bee-trees" are plentiful, and Root (1905) gives detailed instructions for carrying it out.[64] Most well-read modern beekeepers consider that the idea of Bee-hunting is purely American and have no idea that the art is at least nearly two thousand years old.

MANIPULATIONS OF BEE-HIVES

The essential qualifications of a successful beekeeper in Roman times were that he could hive a swarm, take honey from the hives and, when necessary, feed the bees. To hive a swarm was usually a simple operation. It was only necessary to shake the cluster of bees into an empty hive and straightway this became their new home. There was little risk of being badly stung, for swarming bees are exceptionally good-tempered. But to rob the hives of their honey was an entirely different matter, for unless the "vindemiator" were skilled he was likely to receive the stings of a multitude of the infuriated insects. To avoid this risk many of our old "skeppists" first destroyed the bees by means of the fumes of burning sulphur, - a custom which has now almost died out. The modern beekeeper commonly provides himself with a net veil with which to protect his face and sometimes even with gloves for his hands. Skilled apiarists, however, often perform all necessary

---

63 Colum. R.R. IX, 8 and Pallad. V, Tit 7.
64 Root, ABC of Beekeeping, pp. 43-50.] 32

manipulations without any protection for face and hands, relying only on gentle, well-timed movements and careful subduing of the bees.

There appears to be no mention in Latin literature of the use of any special clothing for protection against the stings of the bees nor is there any reference to the revolting system, common in mediaeval and even recent times, of destroying bees in order to take their honey. Indeed all the writers emphasise the need for leaving an adequate store of honey in the hive to sustain the bees during the winter. Thus Pliny quotes Cassius Dionysius who used to leave $1/10$ part to the bees at the summer taking if the hives were full; if less full, a proportionate quantity; and if they were light in weight, or, as he said, "empty", they were not to be touched.[65] Some people, says Pliny, weigh the hives when taking the honey so that they may know how much to leave for the bees, -

"Alvos quidam in eximiendo melle expendunt ita diribentes quantum relinquant".[66]

At the autumn taking, as much as ⅔ of the honey was to be left in the hives and especially that in the combs which contained pollen as well, a very wise precaution, -

"Relinqui ex ea duas partes apibus ratio persuadet et semper eas partes favorum quae habeant erithacen".[67]

Columella recommends that ⅔ be left for the bees at the first harvesting and ⅓ at the later one.[68]

At the present day and in the climate of the British Isles it is commonly agreed that from 20 to 30 lbs of honey are necessary to sustain a stock of bees throughout the winter and spring.

To subdue the bees when taking their honey, smoke was used much in the same way as in modern beekeeping. The "smoker" was a conical earthenware vessel provided with a handle. At the pointed end was a moderate-sized hole from which the smoke issued; at the wider end was a larger hole into which the operator blew with his mouth. The fuel used in this smoker was galbanum or dried cow-dung and when the smoke from this was blown into the hive, the bees, disliking the odour, took refuge in the front, -

"in priorem partem domicilii", or came outside, - "extra vestibulum".[69] The beekeeper could then examine and cut out the combs at his leisure.

---

65   Pliny, H.N. XI. 13.
66   Pliny, H.N. XI. 16.
67   Pliny, H.N. XI. 16.
68   Colum. R.R. IX. 15. 6.
69   Colum. R.R. IX. 15. 6; Pallad. R.R. VII, Tit. 7.

This method of smoking depended for its efficacy on the intimidation of the bees. When the smoke was sufficient to cause them to come out of their hive it was excessive and likely not only to kill some of the bees, - "Fumo nimio inficiuntur",[70] but also to taint the honey. The modern beekeeper administers only sufficient smoke to alarm the bees. This causes them to gorge themselves with honey from the cells, after which they become good-tempered and tractable.

After smoking, cold water was sometimes used to cool the smoked and heated bees, -

> "Suffitas deinde et aestuantes apes refrigerare oportet, conspersis vacuis aqua infusa".[71]

That the honey was ready for harvesting was indicated by fullness of the hive, the great number of the bees and the presence of sealed honey cells and, in the late summer, by the expulsion of the drones.[72]

Columella advises that honey be taken early in the morning, for it is not good that the exasperated bees be provoked in the heat of the day, - "neque enim convenit aestu medio exasperates apes lacessiri":[73] but this is a fallacy, for bees are more irritable when handled in the morning than in the middle of the day when many of the older ones are engaged in foraging.

Varro advises that the honey be not taken in sight of the bees lest they lose heart, -

> " … neque palam facere oportet ne deficiant animum";[74]

but this assumption is quite unwarranted for they are not sufficiently intelligent to know their honey is being taken and they greatly enjoy drinking up the honey dripping from the cut combs.

As previously stated, bees store their honey mostly in the back and upper portions of their dwellings. At the end of each honey season, therefore, the hives were opened and the rearmost honey combs taken out. In the following year, the spaces so made would again be filled with newly-made combs, which accounts for Pliny's statement that the newest combs are chiefly filled with honey, -

> "Novissimi (favi) maxume implentur melle",[75]

---

70   Pliny, H.N. XI, 16.
71   Colum. R.R. IX, 14, 7.
72   Varro, R.R. III, 16, 32 & Colum. R.R. IX, 15, 4.
73   Colum. R.R. IX, 15. 4.
74   Varro, R.R. III, 16, 34.
75   Pliny, H.N. XI, 10.

and that in order that these may be abstracted the hives are turned on end, -

"aversa alvo favi eximuntur".[76]

Columella warns his readers against taking honey from half-filled combs, - "semipleni favi", - but the combs might be cut out when the cells were filled and sealed over with waxen cappings, -

"liquore completi et superpositis ceris tamquam operculis obliti demetantur".[77]

This is excellent and necessary advice even for the modern bee-man who is often tempted to extract unsealed honey, only to find that it is liable to ferment owing to its watery consistency. Before the cells are sealed by the bees the honey is reduced to a proper density by evaporation caused by the heat of the hive.

Varro was under the erroneous impression that the bees were more assiduous and profitable if they were not robbed at all, or but very little, in some years, his idea being that, like land under fallow, they profited from a rest, -

"Ut in aratis, qui faciunt restibiles segetes, plus tollunt frumenti ex intervallis; sic in alvis, si non quotannis eximas, aut non quoque multum, et magis his assiduas habeas apes et magis fructuosas".[78]

But any apparent advantage so obtained would not be due to an improvement in the bees but to the fact that the hives would be well stored with food for winter and likely to send off early swarms in the spring for lack of breeding space.

For cutting the combs from the hives, two special tools, each one and a half feet or a little more in length, were used. The one was a long two-edged knife, of which the end was curved after the manner of a bill-hook so as to form a sort of scraper, -

"culter oblongus ex utraque parte acie lata, uno capite aduncum habens scalprum" and the other was flattened and keenly sharpened at the end, -

"alterum prima fronte planum et acutissimum".[79]

76	Pliny, H.N. XI, 10.
77	Colum. R.R. IX.15.1.1.
78	Varro, R.R. III, 16, 33.
79	Colum. R.R. IX, 15, 4.

Roman Bee Lore - L. E. Snelgrove

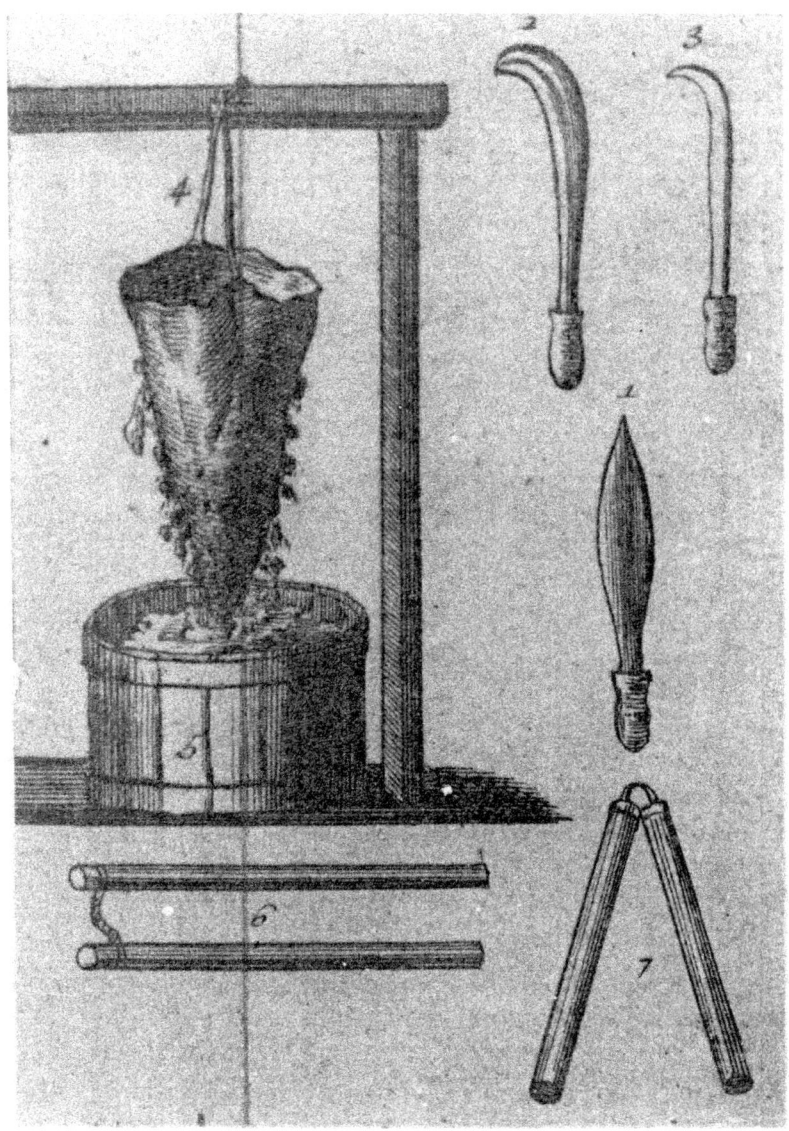

ANCIENT BEEKEEPING APPARATUS

From an illustration in J. Simon's "Le Gouvèrnement Admirable, ou La République des Abeilles", 1758. The knives and straining apparatus are exactly similar to those described by Columella.

"Duobus ferramentis ad hunc usum opus est, sesquipedali, vel paulo ampliore mensura factis, quorum alterum sit culter oblongus ex utraque parte acie lata, uno capite aduncum habens scalprum; alterum prima fronte planum et acutissimum; quo melius hoc favi succidantur, illo eradantur, et quidquid sordidum deciderit attrahatur"

Colum. R.R. IX, 15, 4.

"Saligneus qualus, vel tenui vimine rarius contextus saccus, inversae metae similis, ... obscuro loco suspenditur; in eum deinde carptim congeruntur favi. Deinde ubi liquatum mel in subjectum alveum defluxit, transfertur in vasa fictilia."

Colum. R.R. IX, 15, 13.

The former was used to cut away combs built transversely and pieces adhering to the hive-walls and for hooking debris from the floors; the latter served for cutting down the more regularly built and easily accessible combs, which were received with both arms, - "subjectis duobus brachiis", and so carried away.[80]

Simon (1758) describes the knives used in his time for taking honey as follows, -

(1) "Un grand couteau en forme de tranchelard long, ou dont on se sert pour tuer les pores gras, et pour les saigner; il doit couper parfaitement bien afin de ne pas gâter, flétrir, ni déchirer les raïons de cire."[81]

(2) "Le second outil est une lame plate, large d'un pouce, recourbée par le bout en forme de langue de carpe ... qui sert à passer entre les gateaux de miel pour les détacher au fond de la ruche sans les briser, et avec quoi on les y décole de chaque coté."[82]

These tools, illustrated by Simon, correspond with those described by Columella and were doubtless similar in size, shape and manner of use to those employed by the Romans.

In order to guard against dirt and disease, the modern beekeeper renews the combs in a hive from time to time, removing those that are the worst shaped and contain drone cells, and substituting for them new sheets of waxen "foundation", which the bees soon elaborate into combs. Columella advises that the old and bad combs be cut out when the honey is being taken and that only combs containing honey and brood be left.[83] He further recommends that where combs are built transversely within the entrance, the hive be reversed so that the transverse combs will be harvested when filled with honey in the following year; in this way the bees will be compelled to build new ones, and old combs will be avoided, for, he says,

---

80	Colum. R.R. IX, 15.4.
81	Simon, p. 338.
82	Simon, p. 340.
83	Simon, p 338.

"tanto deteriores sunt, quanto vestustiores".[84]

Varro would cut out empty as well as dirty combs at the honey-taking and Palladius and Didymus are of the same opinion, -

"siqua pars nihil habet, aut habet inquinatum cutello praesecatur".[85]

Didymus gives a curious and fallacious reason for cutting out the worst of the empty combs when spring-cleaning the hives, -

"Si vero multi favi in vasis fuerint deteriores eximere oportet, ne ob loci angustiam aegrotent".[86]

But bees do not sicken from want of space and no modern beekeeper would be so unwise as to cut out combs merely because they were empty, for not only would the bees have to expend unnecessary labour in renewing them, but the new combs would most likely be largely composed of drone-cells.

Aristotle says that beekeepers cut out parts of the combs containing drone cells,[87] which, for the same reason, must have been a futile proceeding.

There were three recognised times for taking honey from the hives. Varro gives them as

(1) At the rising of the Pleiads, - "tempus Vergiliarum exortu", - about June 25th. At this time honey gathered from the fruit blooms would be ready.

(2) At the end of summer, - "aestate acta, antequam exoriatur Arcturus", - about September 12th, when the main crop of honey would have been gathered.

(3) After the setting of the Vergiliae, - "post Vergiliarum occasum", - about November 11th, at which time honey from the heather and late flowering plants would be ready to be taken.[88]

Vergil mentions the first and last of these times for harvesting[89] and Columella only the first and second.[90] Didymus gives all three, but wisely adds that the honey should not be taken on fixed dates but when the honey combs are filled, -

---

84	Colum. R.R. IX, 15, 11.
85	Varro, R.R. III, 16. 35.
86	Geop. XV. 4.
87	Arist. IX, 40. 8.
88	Varro, R.R. III. 16. 34.
89	Vergil, Georg. IV, 231-3.
90	Colum. R.R. IX. 14.5 and IX, 14.7.

"Non tamen definitis et certis diebus sed juxta favorum perfectionem".[91]

With this reservation, the dates given correspond roughly with those on which the honey is harvested in our own times.

All beekeepers know how keenly bees pursue and eat the honey of which they have been despoiled. Columella draws attention to this and advises that all openings in the walls and windows of the honey-house be carefully stopped up, -

"linendaque sunt diligenter, foramina parietum et fenestrarum, nequid sit apibus pervium".[92]

Some beekeepers were sufficiently skilled to take honey without the aid of smoke and this honey known as "ἄκαπνος" was considered the most suitable to be preserved.[93]

Next to the taking of the honey, perhaps the most important duty of the apiarist is to feed the bees, either for the purpose of stimulating the raising of brood, or to supplement the winter stores. The ancients fed only for the latter purpose, but being without sugar they were greatly handicapped in comparison with present-day beekeepers.

All the writers lay stress on the need for leaving an adequate supply of honey in the hives in the autumn, although it was recognised that some hives would not have sufficient even if none were taken from them.

It was commonly known that when bees were starved, they either died in their hives or abandoned them, -

"Inopia cibi desperant moriunturque aut diffugiunt",[94]

the latter probably being a reference to the little-known phenomenon of "hunger-swarming".

An effective, but not over-desirable method of feeding is described by Varro. Vessels were filled with honey and water, - "aqua mulsum" - and into these was placed clean wool, - "laxam puram" - upon which the bees could stand while they drank. One such vessel was placed near each hive and kept filled. This corresponded with the practice known today as open-air feeding. It is seldom adopted because of the possibility of spreading disease from one stock to another and because it sets up robbing. In fact nothing could be better calculated to encourage the latter than the plan of putting an open-air feeder near each hive.

91	Geop. XV, 5.
92	Colum. R.R. IX. 15, 10.
93	Pliny, H.N. XI, 16.
94	Pliny, H.N. XI, 14.

As an alternative to honey, ripe figs were used. They were boiled with water and made into cakes, - "coctas in offas" - which were placed near the bees,[95] or they were bruised and soaked in water or wine and the resulting liquid placed in little troughs, - "canaliculos" - provided with wool and placed near the entrances.[96]

Some people, according to Varro, made cakes of raisins and figs pounded together and mixed with a grape juice boiled down to a third ("sapa"). These they placed in a convenient place where the bees could feed on them when they went out in the winter.[97] But such a plan would be useless, for during the winter bees are seldom willing to take artificial food and it is very doubtful if they would partake liberally of such a mixture at any time.

According to Pliny, carded wool soaked in grape juice, raisin wine, or honied water and even the raw flesh of hens were offered to the bees for food, -

> "Item lanas tractas madentes passo, aut defruto, aut aqua mulsa. Gallinarum etiam crudas carnes".[98]

Didymus prescribes raisin cakes containing a little of the herb savory ("thymbra") for feeding starving bees.

We may safely surmise that the Romans were not very successful in feeding their bees, however, for not only would the bees refuse to eat some of the foods offered to them but most of these, if partaken of, would ferment and produce sickness.

We still meet occasional cottagers who attempt to feed their bees by means of hollowed sticks which, like the "canaliculi" of Columella conduct small quantities of sugar and water into the hives. The method is picturesque as being a survival of a very ancient practice, but is of little or no value.

The Romans occasionally experienced their bad seasons and sometimes had to feed their bees in summer when continued dry weather had deprived them of food from the flowers.

> "Quibusdam etiam aestatibus iidem cibi praestandi, cum siccitas continua florum alimentum abstulit".[99]

95	Varro, R.R. III. 16. 28.
96	Colum. R.R. IX, 14. 15.
97	Varro, R.R. III, 16. 28
98	Pliny, H.N. XXI, 48.
99	Pliny, H.N. XXI, 48.

In our climate, bad seasons are usually due to excessive rain, as, for example, in the summer of 1920, when bees had to be fed almost continuously in some districts to keep them alive.

The Roman beekeepers were enjoined not to disturb their hives during the winter,

"Per tempore hiemis non expedit movere aut patefacere vasa".[100]

A very sound injunction for enthusiastic novices in apiculture.

The Romans sometimes practised operations which are even now attempted only by skilled apiarists.

They knew that the bees never forsake their leader and they found that if he were deprived of his wings, or these were shortened by cutting, he could not fly and therefore a swarm issuing without being noticed would not be lost, -

"ubicumque ille consedit ibi cunctarum castra sunt".[101]

All the Latin writers on bees after Varro recommend this operation on the "King's" wings. Vergil advises "regibus alas eripe",[102] which suggests that the wings were to be plucked out - a cruel proceeding; Columella's words "detractis alis" convey the same idea; Pliny has "Si quis alam ei detruncet," which possibly means that only one wing was shortened by cutting; Palladius uses the words "exectis alis" which imply cutting with a sharp instrument. Didymus however gives the best directions, for he advises that only the ends of the wings be cut off.[103]

The clipping of the queen's wings demands considerable skill, for her delicate abdomen is easily injured and her power of laying eggs destroyed by injudicious handling.

A queen with clipped wings emerges with the swarm and makes an effort to fly, but falls to the ground, where the bees join her. The ancient writers however erroneously believed that she did not come out but remained within the hive.

"Qui destitutus praesidio finem regni non audet excedere",[104] and

"Ipso enim intus manente non discedent".[105]

100	Colum. R.R. IX. 14. 13.
101	Pliny, XI, 17.
102	Vergil, Georg. IV. 103.
103	Geop. XV. 4.
104	Colum. IX, 10, 3.
105	Geop. XV, 4.

Although the clipping of a queen's wings is sometimes recommended, the practice is far from general, for the bees sometimes lose their disabled mother when swarming and return to the hive without her. Even if she is not ultimately lost, they grow tired of a wingless queen and usually rear another in her stead.

If stocks are weak in the spring, they are often united. This involves a difficult operation and as a rule skeppists and other old-fashioned beekeepers do not attempt it. Even with modern movable frame hives, considerable trouble and skill are necessary to cause different stocks to unite peaceably.

Both Columella and Palladius, however, give fairly sound directions for joining weak colonies together.[106]

When a hive contained few bees, its "King" was first to be killed. The stock was then to be replenished by union with a swarm, of which also the "King" was previously killed. But this was to be done, says Columella, only in the spring and when the hive contained some brood. "Cum primo vere in eo vase nata est pullities", a wise precaution, for the bees would then be in a position to raise a new "King". If the hives contained no brood, the populations of two or three were to be joined together and to accomplish this the beekeeper was told first to sprinkle the bees with sweet liquor, - "prius respersas dulci liquore," - and then to keep them shut in a hive for three days, after providing them with honey, -

"apes atque inclusas per triduum tenebimus apposito cibo mellis"

leaving only small holes in the hive for ventilation, -

"exigua tantum spiracula relinquemus in colla."

Neither writer gives details as to how the bees were to be got out of their original hives nor do they say whether their combs were to be utilised, but the directions given are most interesting, for the sprinkling with sweetened liquor corresponds to our modern practice of spraying bees with scented liquids before uniting them in order that they shall not fight; and their imprisonment for three days was sufficient to destroy the memory of their old home and to cause them to accept any new location assigned to them.

Varro remarks that fighting bees should be sprinkled with honied water, for not only do they cease fighting, but they cluster and lick one another, -

"Conferciunt se lingentes"

---

106   Colum. R.R. IX, 11, 1, & Pallad. VII, TIT 7.

and he adds that if wine be used, they apply themselves the more eagerly, and drinking, become stupefied:

"obstupescunt potantes".[107]

Owing to the mistaken belief that the bees of an old "King" would be inclined not to obey those of a younger one and would therefore be punished and put to death, Columella advised the removal of the young rather than the old "King" when uniting stocks, - "Sunt qui seniorem potius regem submoveant, quod est contrarium".[108]

Today old and worn-out queens are discarded and young vigorous ones are retained.

Another advanced feature of Roman apiculture was that of "Requeening," as it is now called. In these days, when the majority of beekeepers are unable safely to introduce a new queen to a stock without much help and advice, it is astonishing to find that both Varro and Columella speak of requeening as an ordinary matter. The former says that the less vigorous bees were put away and placed under a new King, -

"imbelliciores secretas subiiciunt sub alteram regem"[109]

- and Columella says that when the "King" is dead, a successor is chosen from another colony having several, transferred to the kingless bees and set up as their ruler, -

"domino mortus ... ex iis alvis quae plures habent principes, dux unus eligitur; isque translatus ad eas quae sine imperio sunt, rector constituitur."[110]

The breeding and introduction of queens is now a highly technical branch of apiculture which demands the highest skill on the part of the beekeeper. We are not told what devices the Romans used to secure the safe introduction of the queens, but it is none the less surprising that they practised what are generally regarded as peculiarly modern expedients for maintaining the welfare of their bees.

The ancients also knew the value of adding ripe brood to weak colonies to strengthen them. Both Columella and his faithful imitator Palladius recommend that when stocks have been weakened by disease, the bee-master should cut out combs containing ripe brood and if possible some queen cells as well, -

"cerae, quae semina pullorum continent partem recidere, in quae regii generis proles animatur".[111]

107	Varro, R.R. III, 16. 35.
108	Colum. R.R. IX, 11. 2.
109	Varro, R.R. III, 16. 35.
110	Colum. R.R. IX, 11. 3.
111	Colum. R.R. IX, 11. 4.

These combs were not to contain unripe brood, which would perish from exposure, -

> "Si immaturos transtuleris, interibunt"[112]

but the ones chosen were to contain hatching bees, -

> "quando erosis cooperculis ad nascendum maturi (pulli) capita nituntur exerere",[113]

all of which is thoroughly sound, except that the weak stocks of bees which received the additions of brood would in most cases refuse to accept queen-cells as well and would immediately break them down.

The strengthening of weak hives by adding brood from others is quite easy with modern moveable comb-hives, and subject to reservations respecting the spreading of disease, is freely practised by modern bee-men. The fixed-comb hives used by the Romans, however, made the processes of cutting out and introducing brood combs very difficult and there is no wonder that some of the operations, such as this, described by the Latin writers, fell into desuetude during mediaeval and even comparatively modern times.

It is commonly recognised that swarms of bees vary in respect of prolificness and working qualities, and during the last fifty years efforts have been made to breed by selection so as to obtain improved strains. The chief obstacle to this is the difficulty of controlling the fertilisation of the queens, for crossing is inevitable if a non-selected stock of bees exists within three or four miles of the home of the young queens reared.

It is remarkable that Columella appears to have been aware of the difference in strains of bees and the results of undesirable crossing, although he could not have had any idea as to how the latter was effected. He distinguishes bees as good and bad, - "bonae" and "improbae" and points out that there is the same trouble and expense in keeping both. He therefore gives a special warning to those who would acquire bees not to mix the degenerate with the nobly-born, for the latter are thereby defamed and the crop of honey is diminished, -

> "quod maxime refert, non sunt degeneres intermiscendae, quae infament generosas, nam minor fructus mellis respondet cum segniora interveniunt examina".[114]

---

112   Pallad. R.R. VII, Tit. 7.
113   Pallad. R.R. VII, Tit. 7.
114   Pallad. R.R. VII, Tit. 7.

5
HONEY

ORIGIN OF HONEY

It is popularly supposed that bees gather honey from the nectaries of flowers and convey it to their hives where they deposit it in the cells of combs for future use as food. This is not quite accurate, however, for the sweet substance secreted by the flowers is a watery solution of cane-sugar, commonly termed "nectar", which does not become honey until it has been mostly inverted by a saliva produced by the bee in the act of swallowing. After inversion it consists of approximately equal parts of two forms of grape sugar, dextrose and levulose, with a small amount of unchanged cane sugar and some water. The nectar does not enter the bee's stomach but is carried in a small sac, - the "honey sac" - whence it is regurgitated and deposited in the cells of the combs. As the inversion of cane sugar is equivalent to digestion, honey is an easily assimilable and therefore highly prized food.

"Nectar" is produced not only by flowers but sometimes also by the stipules and leaves of certain plants. Certain kinds of aphides, the objectionable plant lice which infest the leaves of trees and shrubs in the summer, secrete and extrude from their bodies a sugary solution which, deposited on the leaves, is sometimes collected by bees and is known to bee-keepers as "honey-dew". As this usually contains much gummy matter, the ordinary honey is spoilt by being mixed with it.

The true nature and origin of honey remained a mystery to the ancients. That it was gathered from flowers and sometimes from leaves was commonly recognised. Aristotle considered that the bees did not make honey, but only collected that which fell, and he argued that if it originated in flowers the bees could replenish their stores in autumn when there are still plenty of flowers.[1] His difficulty is accounted for by the fact that the flowers secrete nectar only when the temperature is comparatively high, and the bees cannot therefore obtain much honey in autumn.

Vergil speaks of the celestial gifts of serial honey, -

"Aerii mellis coelestia dona",[2]

and Pliny, too, considers that it is an exudation from the air, -

---

1   Arist. H.A. V, 22, 4.
2   Vergil, Georg. IV, 1.

"Venit hoc (mel) ex aere".[3]

The same writer refers to other picturesque beliefs concerning its origin. Some thought it was a sweat of the sky, - "caeli sudo"; others that it was a saliva coming from the stars, - "quaedam siderum saliva"; and others that it was a juice coming from the air as it purged itself, - "purgantis se aeris succus".[4]

Seneca, however, discussing the origin of honey, comes remarkably near the truth. He says that it is not sufficiently established whether the bees draw a juice from the flowers, which straightway becomes honey, -

"Non satis constat, utrum succum ex floribus ducant qui protinus mel fit",

or whether what they have collected they change into honey by some mixture and quality of their breath, -

"an quae collegerunt in hunc saporem mixtura quadam et proprietate spiritus sui, mutent".

Some people, he says, consider that the bees have not the ability (scientiam) to make honey but only to collect it, while others believe that they do not turn what they gather into honey without the aid of some ferment, so to speak, -

"Quidam existumant ... in hanc qualitatem verti ... non sine quodam, ut ita dicam, ferment,"[5]

- a most intelligent anticipation of what was not definitely ascertained until nearly 1800 years after Seneca's time.

Aristotle had noticed that bees ate their honey when they were smoked, which they were not seen to do at other times, and he believed that they stored it up for food.[6] But it was not considered to be their ordinary food, for Pliny says that an abundance brings laziness and that then the bees feed on honey instead of pollen, -

"Copia (cibi) ignaviam adfert ac jam melle non erithace pascuntur".[7]

The truth is that bees always eat what they need whether of honey or pollen. The apparent laziness might be caused by the overloading of the hive with an excess of honey which would occupy the space needed for brood-rearing.

---

3   Pliny, H.N. XI, 12.
4   Pliny, H.N. XI, 12.
5   Seneca, Epist. LXXXIV.
6   Arist. H.A. IX, 40. 2.
7   Pliny, H.N. XI, 14.

PROPERTIES AND CARE OF HONEY

When the honey combs had been taken from the hives, the portions containing pollen and brood were first cut away, - for otherwise, as in the case of skep honey of to-day, these gave a bad taste to the honey, -

"malo sapore mella corrumpant",

- and, according to Palladius, the remainder were then placed in a very clean cloth, - "nundissimum sabanum" and the honey pressed through into a vessel placed to receive it.

The need for ripening new honey - "mel recens" - was not forgotten and as at the present time, it was to be left in the vessel for a few days and the scum which rose to the top was to be skimmed off.[8]

Pliny correctly remarks that the new honey was as though diluted with water, -

"Est autem initio mel ut aqua dilutum".[9]

We now know that it actually is diluted with water, the excess of which evaporates during the ripening, leaving the honey of a thicker consistency, - which, says Pliny, is attained after 20 days, - "vicessimo die crassescit".

Palladius considered that the best honey was that which flowed through the cloth without pressure and provided it were not unripe this would be true. The pressing of the combs would expel pollen as well as honey from the cells and so the flavour would be affected.

Columella gives more elaborate instructions for straining the honey. It should be done, he says, while the combs are still warm, - "eodem die, dum tepent"; these should be placed bit by bit ("carptim") in a sort of straining basket or sack made of loosely woven twigs shaped like an inverted cone, - "saligneus qualus vel tenui vimine rarius contextus saccus inversae metae similis"; after the portions containing brood ("pullos") or pollen ("rubras sordes") had been removed, the basket was hung up in an obscure place where the bees would not find it; the liquid honey flowed into a tub ("alvum") placed underneath and was then transferred into earthenware vessels ("vasa fictilia") which were to be left open for a few days and the honey frequently skimmed with a ladle, - "saepius ligula purgandus"; when it had mostly run through, the pieces of comb were removed from the strainer and from them a second-grade honey was obtained by pressure; this was kept apart ("seorsum reponitur") by careful beekeepers lest it should spoil the best flavoured, -

---

8   Pallad. R.R. VII, Tit 7.
9   Pliny, H.N. XI, 13.

"ne quod est optimi saporis hoc adhibito, fiat deterius".[10]

These instructions could hardly be improved upon and the beekeeper who scrupulously followed them was considerably in advance of most of our present-day skeppists. It is specially interesting to observe that Simon (1758) illustrates and recommends for use almost the identical apparatus described by Columella. (see illustration on p. 102).

Some of the Roman writers were well acquainted with the characteristic qualities of good honey, - so much so, indeed, that certain of their tests would be sufficient for the guidance of a judge at a modern honey show. Thus Pliny gives as evidence of good honey, that it have aroma ("ut sit odoratum") and that it be sweet yet pungent to the taste ("ex dulci acre"), of good consistency ("glutinosum") and clear ("perlucidum"). A test for good consistency was that on being touched it could be drawn out in slender threads, -

"tactu praetenuia fila miltit".

But if the threads were at once broken and the drops fell back, the honey was considered to be inferior.[11]

Diophanes commends honey that is clear ("pellucidum"), pale yellow ("subflavum"), smooth to the touch ("tactu laeve"), tenacious ("sibiipsi cohaerens et continuun"), slow-flowing ("sublatum in sublime sensim defluat"), thick ("crassun") and having a good odour ("odor bonus").[12]

Most kinds of honey solidify or "candy" on keeping, the rapidity with which this happens depending largely on the proportions of the contained dextrose and levulose. If the latter is in excess, the candying will be slow and not complete. Candying is often regarded as a sign of purity but this is fallacious. Pliny tells us that the honey which candies is least esteemed, -

"Quod concrescit autem minume laudatur".[13]

At the present day, honey is often adulterated by unscrupulous dealers, and people doubtless consider this practice to be a quite modern development due to competition in trade. But "there is nothing new under the sun" and so we find that even the ancients sometimes attempted to make or adulterate honey. Herodotus speaks of certain people of Asia Minor who made a kind of honey from the flowers of the heaths and flour, though how this was done it is difficult to imagine. Speaking of the Zygantes, he says they were

---

10	Colum. R.R. IX, 15, 12 & 13.
11	Pliny, H.N. XI, 15.
12	Geop. XV, 7.
13	Pliny, H.N. XI, 15.

believed to make more honey than their bees.[14] Pliny tells us that grape must was boiled down for the adulteration of honey and Diophanes says that if pure and adulterated honies be immersed in water, that which is not adulterated will not soil the hands, -

"Quod enim adulteratum non est, tangentis manus non faedabit".[15]

KINDS OF HONEY

Pliny says that the most commendable honey was the golden red ("rutilum") such as would be made in dry weather. The most agreeable ("aptissimum") was that of thyme, which was of a golden colour and very pleasant to the taste, - "coloris aurei saporis gratissimi"; it did not candy and was viscid and of high density.[16] Diophanes agrees that the attic (thyme) honey does not candy but he adds that it darkens in colour, -

"Atticum (mel) liquidum permanet colore autem nigrescit",[17]

and this is rather borne out by Mahaffy[18] who in his "Rambles and Studies in Greece" says, -

> "This honey of Hymettus, which was our daily food at Athens, is now not very remarkable either for colour or flavour. It is very dark and not by any means so good as the honey produced in other parts of Greece."

According to Morley, who quotes the above passage, there are to-day as many hives as persons in the province of Attica.

The thyme honey of Hymettus was the most famous in the world and was rivalled only by that produced at Hybla in Sicily, - probably Hybla Minor of Heroea in the southwest of the island,[19] - this honey also being gathered from thyme. There are numerous references in Latin literature to the honey and bees of these famous districts. Vergil for example speaks of a willow hedge pastured by Hyblaean bees, -

" ... saepes

Hyblaeis apibus florem depasta salicti",[20]

---

14 Herod. VII, 31 & IV, 194.
15 Geop. XV, 6.
16 Pliny, H.N. XI, 15.
17 Geop. XV, 7.
18 Cited by Morley, p.374.
19 De Soignie, p. 95.
20 Vergil, Ecl. 1, 54-5.

and Ausonius has a similar passage,

"Hyblaeis apibus saepes depasta susurrat".[21]

Ovid in his laments says, -

"Et careat dulci Trinacris Hybla thymo",[22]

and Martial in one of his famous epigrams considers himself more honoured because his works are widely read than if Hymettus or Hybla fed his bees, -,

" ... si ...

Pascat et Hyblas meas, pascat Hymettos apes."[23]

He calls the bee the plunderer of Thesean Hymettus and speaks of the noble nectar gathered from the Palladian woods, -

"Hoc tibi Thesei populatrix misit Hymetti Pallados a sylvis nobile nectar apis."[24]

In another epigram Martial states that the Hyblaean honey might be mistaken for that of Hymettus, -

"Cum dederis Siculos mediis de collibus Hyblae

Cecropios dicas tu licet esse favos",[25]

and in yet another he asks how he can be expected to make lively epigrams on dull subjects, which is like trying to obtain Hyblean or Hymettan honey by offering to the Attic bee only Corsican thyme, -

"Mella jubes Hyblaea tibi vel Hymettia nasci;

Et thyma Cecropiae Corsica ponis api",[26]

Corsican honey being well-known for its inferior quality.

In his "Bellum Puncicum", Silius Italicus represents Hybla as challenging Hymettus with its nectared combs, -

"Tum quae nectareis vocat ad certamen Hymetton Audax Hybla favis ... ".[27]

---

21 Auson. Epist. XXV, 12.
22 Ovid, Tristia V, 13, 22.
23 Martial, Epig. VII, 87.
24 Martial, Epig. XIII, 104.
25 Martial, Epig. XIII, 105.
26 Martial, Epig. XI, 42.
27 Sil. Ital. Punicorum XIV, 199-200.

Another thick honey was obtained from rosemary, -

"Marino e rore spissum".[28]

It is from this plant that the famous honey of Narbonne is gathered. Until a few years ago this was perhaps the most renowned honey of modern times, but unfortunately the beekeeping of the district has largely given place to the cultivation of the vine and the supply of the famous French honey has diminished.

The queen of nectiferous plants is the common white clover (Trifolium repens) and wherever this abounds, large crops of excellent honey are often obtained. The ancients could not have realised the value of this flower for no mention is made of its honey by any of their writers. It is nevertheless recognised as an important source of honey in Italy today[29] as well as in many other countries which formed parts of the Roman Empire.

The worst honey in common use came from Corsica and Sardinia. It was disliked on account of its bitterness which was variously ascribed to different plants. Dioscorides says it was due to pasturage on absinthe (Artemisia Absinthium) or the wormwood plant, -

"τό δέ ἐν Σαρδωνίᾳ γεννώμενον μέλι πικρόν διά τήν τοῦ ἀφινθίου νομήν".[30]

Vergil makes Lycidas say to Moeris, - "Sic tua Cyrneas fugiant examina taxos",

and Forbiger in notes on this passage says:

"Taxus omnino mellificio exitiosa quia mel amarum inde fiebat," and

"Mel amarum autem propter multas taxos in Corsica insulam confectum satis est notum."[31]

Ovid, apostrophising the disappointing waxen tablets received from his mistress Corinna, says he thinks the wax was collected by the Corsican bee from the flower of the tall hemlock under its infamous honey, -

"Quam, puto, de longae collectam flore cicutae

Melle sub infami Corsica misit apis",[32]

and Diodorus Sieulus alleges that the bitter flavour of Corsican honey was due to the box trees on the island, -

---

28 Pliny, H.N. XI, 15.
29 Benussi Bossi, p. 192.
30 Dioscor. 11, 102.
31 Vergil, Ecl. IX, 30.
32 Ovid, Amor. 1, 12, 10.

> "Multum quoque buxi, ejusque non vulgaris, ibi nascitur: Quae causa est ut mel prorsus amarum illis existat."[33]

Pliny considered that the Corsican wax had medicinal value since it was made from box, -

> "Post has Corsica (cera), quoniam ex buxo fit, habere quandam vim medicaminis putatur".[34]

The present writer can find no evidence that honey is gathered in any quantity from any of these trees or plants or that they embittered the Corsican honey. M. Bourgeois, in a series of articles on "Beekeeping in Corsica" (1918-1920), says, - "Corsican honey is of very fine quality and aroma when it is collected in the spring; the autumn honey has however a strong bitter taste and is dark in colour." He adds that a "reddish bitter honey" is gathered from the chestnut tree; "much dark honey with a strong flavour" from the false pepper plant; and "much light-coloured honey in summer, reddish and thick in the autumn" is gathered from the heather. Amongst an exhaustive list of Corsican honey-bearing plants he does not give absinthe, yew, hemlock, or box.[35]

Heather honey has a strong flavour which is disagreeable to many people and it was probably this which caused the Romans to regard the Corsican honey in general as inferior. Indeed, Pliny affirms that heather honey was the least approved, -

> "Tertium genus mellis minumae probatur silvestre quod ericaceum vocant".[36]

It seems likely that yews were deemed to be obnoxious to bees because the leaves are poisonous to cattle, and the bitter taste of wormwood and poisonous qualities of hemlock may account for the bitter taste of the Corsican honey being erroneously attributed to them. It by no means follows that the bitter or poisonous qualities of plants are carried in the honey gathered from them.

Thus Deane, referring to fields of the poisonous Atropa Belladonna at Long Melford says, "No bad results from eating the honey ... have been reported."[37]

A number of plants, all belonging to the Ericaceae, or Heath family, are considered to yield honey which is poisonous to man. Amongst these are the Kalmia latifolia, Rhododendron maximum and Azalea hirsuta.[38]

---

33 Diod. Sic. V, 14.
34 Pliny, H.N. XXI, 49.
35 Bourgeois, in L'Apiculteur. Cited in Journ. Agric. 1920, p. 882.
36 Pliny, H.N. XI, 16.
37 Deane, B.B.J. Vol. XLI, 358.
38 Morley, p. 193.

In his account of the celebrated retreat of the ten thousand, Xenophon relates how many of his soldiers were overcome by the honey they found near Trebizond in Asia Minor. "In those villages", he says "were great quantities of bee-hives. All the soldiers who ate of the honey lost their senses and were seized with a vomiting and purging, none of them being able to stand upon their legs. Those who ate but little were like men very drunk, and those who ate much like madmen, and some like dying persons. In this condition great numbers lay upon the ground as if there had been a defeat and the sorrow was general. The next day none of them died but recovered their senses about the same hour they were seized, and the third and fourth day they got up as it they had taken physic".[39]

Deane, citing Dr. Thresh in a paper read to the Pharmaceutical Society, writes, -

"Mr A Billiotti, H. M. Consul at Trebizond in his Report for 1879 says that 'bees are reared on a somewhat large scale in the province of Trebizond, but the honey produced is unfit for food … It is presumed that the poisonous principle contained in the honey is gathered from the flowers of the Datura Stramonium which grows in abundance on the coasts. Beehives, therefore, are remunerative only for their wax.' Dr. Thresh examined some of the honey and subsequently concluded that Azalea Pontica and not Datura was the source of the poisonous honey from Trebizond."[40]

Billiotti's assumption was probably another example of attributing the poisonous qualities of a plant to the honey produced by it. The Datura Stramonium flower has such a long corolla tube, - "fleurs de plus de 5 centimetres et demi de longueur",[41] - that hive-bees cannot reach its nectar which is taken by long-tongued moths and even long-billed birds. Bonnier describes it as "venineuse" but does not classify it as a bee-flower.

Pliny states that the honey at Heraclea in Pontus was very pernicious in some years but that writers had not decided what flowers were the cause. There is, however, he explains, a plant called "aegolethron", - "goat's death" - which is fatal to beasts and especially to goats, and of which the flowers in a wet spring conceive a noxious poison. The signs that the honey was poisonous were: that it did not solidify, was redder in colour, had an unnatural odour which provoked sneezing and was denser than harmless honey.[42] Those who ate of it were affected much in the same way as Xenophon's soldiers. They threw themselves on the ground dripping with perspiration and seeking to be cooled, -

---

39   Xenophon, Anab. IV, 8. (Spelman's Trans.).
40   Deane, B.B.J. Vol. XLI, 358.
41   Bonnier, Les Noms des Fleurs, 550.
42   Pliny, H.N. XXI, 44.

"Qui edere, abjiciunt se humi refrigerationem quaerentes; nam et sudore diffluant".[43]

There were many remedies for such poisoning, says Pliny, one of them being old honey-wine made with the best honey and rue, -

"Mulsum vetus e melle optimo et ruta",

or what was probably much more effective, salt meats, if taken so as to induce frequent vomiting, -

"salsamenta etiam, si rejiciantur sumta crebro".

The wine made from the poisonous honey when matured was said to be innocuous; and when the honey itself was mixed with "costus", there was nothing better as an emollient for women's complexions, or when used with aloes for the treatment of bruises, -

"feminarum cutem nullo melius emendari cum costo, sugillata cum aloe".[44]

From Pliny too, we learn that the bees of the Sanni who inhabited a part of Pontus produced honey which caused madness, and which was therefore called "Μαινόμενον". It was with the wax made by these bees that the Sanni paid their tribute to the Romans.[45]

Aelianus makes a rather ludicrous statement about the Pontic honey from Trapesus (Trebizond), which, he says was believed to be made from "box". It was of bad odour and while it restored the health of those who were affected in their minds or seized with falling sickness, it made healthy persons mad, -

"Hocque (mel) esse odoratu grave quod eos qui deducti sunt de mente vel comitiali correpti ad sanitatem restituat; sanos vero dementet".[46]

Democritus, quoting Aristotle, expresses the same idea, -

"Prodit Aristoteles ex buxo mel fieri gravis odoris; ex quo sani edant, mente ipsos moveri, si vero mente perculsi eo vescantur, statim a malo liberari".[47]

There is no doubt that some honeys are noxious to human beings, but their effects were undoubtedly exaggerated by ancient writers. Pliny himself marvels that the bees who carry them in their mouths and fashion them are not thereby killed, -

---

43	Pliny, H.N. XXI, 44.
44	Pliny, H.N. XXI, 44.
45	Pliny, H.N. XXI, 45.
46	Aelian. V, 42.
47	Geop. XV, 9.

"Mirum tamen est venena portantes ore, fingentes ipsas non mori."[48]

The subject of poisonous honeys is even now little understood and Prof. Cook goes so far as to say that he doubts very much whether the honey from any plant is poisonous.[49]

HONEY-DEW

Reference has already been made to "honey-dew", a sweet exudation from leaves or an excretion of aphides. This is gathered freely by bees in hot weather, especially when melliferous flowers diminish in quantity. It is not at all uncommon to find street pavements under trees quite moist with this honey-dew during the months of July and August. The leaves of the oak, lime, sycamore and maple, together with those of many fruit trees, are often so heavily laden with the sweet liquid that it drops on to the ground beneath. The ancients were familiar with its fall but did not understand its origin, and this was doubtless the cause of the common belief that honey fell from the sky or the air. Pliny evidently refers to it when he says that early in the morning the leaves of the trees are found bedewed with honey, and that persons who go out under the open sky early in the morning feel their clothes and hair stuck together by it, -

"Prima aurora folia arborum melle roscida inveniuntur ac si qui matutino sub divo fuere, unctas liquore vestes capillumque concretum sentiunt".[50]

Some of the commentators have taken Vergil's "pinguem tiliam"[51] to refer to the honey-dew on the leaves of the lime-tree on which, amongst other trees, the poet says the bees forage. But although this tree at times bears large quantities of honey-dew on its leaves, its flowers yield much honey of excellent quality and the term "pinguem" may therefore be appropriately applied to the tree on either account. In his fourth Eclogue, however, Vergil refers definitely to honey-dew which exudes from oak trees, -

"Et durae quercus sudabunt roscida mella".[52]

Aelianus also speaks of the same sweet substance when he says that honey distils from trees in Media, -

"Itaque in Media mel ex arboribus stillare;"

that in Cithaeron, drops flow from the sweet branches, -

48	Pliny, H.N. XXI, 45.
49	Root, A.B.C. p. 263.
50	Pliny, H.N. XI, 12.
51	Vergil, Georg. IV, 183.
52	Vergil, Ecl. IV, 30.

"Cithaerone dulceis ex ramis guttas";

and that in India, especially in the district of the Prasii, it rains liquid honey, -

"In India et maxime in Prasiorum regione liquido melle pluit."[53]

USES OF HONEY

It would be impossible within the scope of this book to notice all the uses to which honey was put during Roman times. As a food it was commonly eaten alone, or as a constituent of cakes, or in conjunction with other dishes. Varro informs us that it was served at the commencement of a banquet and again in the second course, -

"Mel ad principia convivii et in secundam mensam administratur",[54]

and from Athenaeus we learn of many luxurious combinations of foods, as for example, cheesecakes and even young thrushes well steeped in honey.[55]

In his epigram on the "pistor dulciarius" Martial says that the frugal bee appears to work only for the confectioner, - "huic uni parca laborat apis".[56] As in the present day, honey was believed by many to conduce to health and long life. The philosopher Democritus, who lived to be 109 years of age, when asked how men might remain healthy and live long, replied that they should constantly cleanse their bodies with oil and line their interiors with honey, -

"Si externas corporis partes oleo, internas melle illinerent".[57]

Democritus had probably learnt this from the orator Pollio, who at the age of 100 years was the guest of Augustus Caesar and on being asked by him how he had preserved his wonderful vigour of mind and body gave the famous reply, -

"Intus mulso, foris oleo",[58] mulsum being honied wine.

Honey was offered in sacrifices to the gods who patronised agriculture and especially in honour of the goddess Ceres, for whom libations were made with milk, honeycombs and wine, -

---

53	Aelian, V, 42.
54	Varro, R.R. III, 16. 5.
55	Athenaeus, Deipnosoph. 1 & 11.
56	Martial, Epig. XIV, 222.
57	Paxamus, Geop. XV. 6.
58	Pliny, H.N. XXII, 53.

"Cuncta tibi Cererem pubes agrestes adoret Cui tu lacte favos et miti dilue Baccho".[59]

In the worship of Bacchus, who was supposed to enjoy and to have discovered honey, the god was honoured by offerings of hot cakes on which was poured white honey, -

"Melle pater fruitur; liboque infusa calenti Jure repertori candida mella damus."[60]

As a gift few things could be more acceptable than honey and Martial, describing the farm of Faustinus, speaks of the rustic who comes to salute the master, not with empty hands, but carrying a gift of ripe honey in the comb, -

"Nec venit inanis rusticus salutator:

Fert ille ceris cana cum suis mella".[61]

As a medicament, honey was one of the most useful substances in the Roman pharmacopoeia. It was constantly prescribed either alone or combined with other things too numerous to mention. Perhaps one of the most interesting uses to which it was put is that mentioned by Lucretius who says that when physicians wished to give bitter medicine (wormwood) to children, they first smeared the rim of the cup with the sweet and yellow liquid of honey so that the little ones, deluded at their unsuspecting age by this snare for the lips, might drink up the bitter liquid, -

"Sed, velutei puereis absinthia tetra medenteis

Quum dare conantur, prius oras pocula circum,

Contingunt mellis dulci flavoque liquore,

Ut puerorum aetas improvida ludificetur

Labrorum tenus; interea perpotet amarum

Absinthi laticem"[62]

Pliny considered that honey, as distinguished from salt, was a preventive of putrefaction on account of its sweetness, -

"Mellis qualis ipsius natura talis est, ut putrescere corpora non sinat, jucundo sapore atque non asperso, alia quam salis natura".

It was useful in affections of the jaws ("faucibus") and throat ("tonsillis"); for quinsy ("anginae") and all complaints of the mouth ("omnibus oris desideriis") and for the

---

59   Vergil, Georg. 1, 343-4.
60   Ovid, Fasti, 111, 761-2.
61   Martial, Epig. III, 58, 33-4.
62   Lucretius, De Rer. Nat. 1, 937. sqq.

parching of the tongue in fevers ("arescentique in febribus linguae"). As a decoction it was prescribed for pneumonia and pleurisy ("peripneumonicis, pleuriticis"), and it was a remedy for wounds caused by snakes ("vulneribus, a serpenti percussis"). For paralysis it was administered in honied wine ("paralyticis in mulso"). With oil of roses it was injected into the ears ("auribus instillatur cum rosaceo") and it was supposed to kill vermin of the head ("lendes et foeda capitis animalia necat").

On the other hand, continues Pliny, it inflates the stomach ("stomachum inflat"), increases the bile ("bilem auget"), creates nauseousness ("fastidium creat") and when used alone is considered to be useless for the eyes ("oculis per se inutile"), although there are some people, he says, who advise that the corners of the eyes, when ulcerated ("angulos exhulceratos"), be touched with it.[63]

Although there is little or no value in honey when used for some of the purposes mentioned by Pliny, he is much more reasonable than when he discusses many other things used as remedies; and in respect of affections of the mouth and throat, at least, he is justified by the universal use of honey for these in the homely remedies of to-day.

---

63   Pliny, H.N. XXII, 50.

6
WAX

Of the numerous kinds of wax of mineral and vegetable origin now used in industry, none are superior and few are comparable in value to that produced by the bees. From the earliest times its great value was recognised and the Egyptians, Greeks and Romans made use of it in innumerable ways. So indispensable did it become that the Romans preferred it even to money or corn when exacting tributes from certain subject states and in some countries bees were farmed more for the sake of their wax than for the honey they produced.

ORIGIN OF WAX

That beeswax is a secretion from certain glands situated on the underside of the worker bee's abdomen has already been explained. According to Cowan,[1] the resulting wax scales were probably first observed by Martin John in 1684, but it was not until 1793 that Huber began his series of experiments which confirmed the existence and functions of the wax glands. Up to that time it was commonly believed that beeswax was gathered from flowers and trees. This was the opinion of Aristotle, - "Τίγνεται δέ κηρίον μέν ἐξ' ἀνθῶν";[2] amongst the glutinous trees supposed to provide it being the willow, elm and olive, as well as a number of smaller plants. The bees were believed to carry it as they do pollen,[3] having first transferred it from their forefeet to their middle legs and thence to their hind legs.[4]

Pliny too believed that wax was fashioned from the flowers of all kinds of trees and plants except, very curiously, the sorrel and globe-thistle, -

> "Ceres ex omnium arborum satorumque floribus confingunt excepta rumice et echinopode"[5]

and Celsus, cited by Columella, says "ex floribus ceras fieri".[6]

---

1 Wax Craft, p. 48.
2 Aristotle, H.A. V, 22, 4.
3 Arist. H.A. IX, 40, 2 & 3.
4 Arist. H.A. IX, 40, 7.
5 Pliny, H.N. XI, 8.
6 Colum. R.R. IX, 14, 20.

It is interesting to note how, with a little embellishment, the ideas of the ancient writers respecting the origin of wax persisted through the Middle Ages. In 1720 Bradley wrote the following account of it, -

"The Bee gathers both Honey and Wax from the same flowers but not with the same Organs ... As for the Wax, which is the dust of the stamina of the Flowers, the Bees gather it with the two foremost of their Six Legs and convey it into a small Cavity between the two hindmost; They often compress it with their Legs, to the end that they may carry off no small Particles of Wax by Means of those Hairs, wherewith their bodies are covered all over".[7]

WAX RENDERING

The manner of rendering the emptied honeycombs into wax as described by Columella and Palladius was practically the same as that pursued today by beekeepers who are not equipped with "wax-extractors". The pieces of comb, after being rinsed in water ("perlutae") to remove the last traces of honey, were cut into pieces and placed in a metal pot ("in vas aeneum"); they were then covered with water and melted by heat ("adjecta aqua liquantur ignibus") and when that was done the liquid wax was strained through straw or bulrushes (" per stramenta vel juncos defusa colatur"), re-melted and directed into moulds ("in formas") previously moistened with water to prevent the waxen cakes from adhering to them.[8]

Pliny gives some useless additional instructions for the process. He prefers an earthenware pot and advises that the combs, after being washed, be dried for three days in the dark ("triduo in tenebris siccatis"); after being melted in water and strained, the resulting wax was to be re-boiled in the same water ("cum eadem aqua") and the moulds into which it was run were to be smeared round with honey ("vasis melle circumlitis"). There could be no point in drying combs which were to be afterwards placed in water, and as the main difficulty in rendering is to separate the pure wax from the accompanying dross which sinks below and partly adheres to the wax in the water, it would be folly to boil it in the same water a second time. Nor would there be the slightest advantage in lining the moulds with honey.

The best wax was known as Punic, the term not necessarily denoting its origin, for it was made from ordinary yellow wax ("cera fulva") and Pliny gives detailed directions for its preparation. It was distinguished by its purity and whiteness and was considered to be

---

7   Bradley, cited in B.B.J. Vol. XXXIX, 142.

8   Colum. R.R. IX, 16 and Pallad. VII, Tit. VII.

the most useful for medicines, - "Punica medicinis utilissima". To produce it, ordinary wax was first exposed to the open air ("sub divo"), then boiled in deep-sea water to which was added a little saltpetre ("aqua marina ex alto petita addito nitro"). The flower or whitest part of the wax was then skimmed off with ladles and poured into vessels containing a little cold water. It was then again boiled in sea-water alone ("rursus marina decoquunt separatim") and cooled. When this had been repeated three times, the wax was placed on a rush mat ("juncea crate") and exposed to the rays of the sun and moon, the former being supposed to dry ("sol siccat") and the latter to whiten it ("Luna … candorem facit"). In order that the sun might not melt it, the wax was covered with a thin linen cloth ("protegunt tenui linteo"). After being thus exposed it was boiled again, the resulting product being of the greatest whiteness.[9]

The effective features of this treatment were the repeated boilings in water which removed dirt and some of the colour from the wax and the exposure of the purified substances to light, especially direct sunlight, which had the effect of bleaching it. In modern wax-bleaching factories the same processes are used. The crude wax is subjected to repeated melting in slightly acidulated water and separated from the impurities which fall to the bottom of the cake as it solidifies. The purified wax is then drawn out into ribbons which are exposed to the sunlight on strips of canvas for many days. Wax bleached in this way is superior to that whitened by chemical means.[10] Except that the useless sea water and nitre are dispensed with, there is no essential difference between our own method of refining wax and that described by Pliny.

From the same writer we learn that next to Punic wax, that coming from Pontus, where the poisonous honey was produced, was most esteemed; it was of a dark yellow colour ("fulva") and had the odour of honey. Next came the Cretan wax which contained the most propolis and after that the Corsican variety, which, supposed to be gathered from the box-tree, was believed to have some value in medicine.[11]

USES OF WAX

Not only was the whiteſt and pureſt Punic wax in demand, but for innumerable purposes ("innumeros mortalium usus") wax of various colours was required. It was blackened by the addition of papyrus ashes ("chartarum cinere") and reddened by bugloss ("anchusa") or minium (red lead). Variously coloured waxes were used for the production of images

---

9   Pliny, H.N. XXI. 49.
10 Cowan, Wax Craft, PP. 81-8.
11 Pliny, H.N. XXI. 49.

and even walls and armour were sometimes protected from the air by thin coatings of wax.[12]

The Romans were much addicted to the making of images and for this purpose they were skilled in the use of various metals, stone and wax. Pliny, satirising a lapse of the popular taste in the art of imagery in his day, says that in earlier times ancestors kept in their halls models or images of their forefathers, fashioned, not of brass or marble, but of wax and showing a likeness to the individuals they represented. When a member of a noble family died, the funeral procession included not only the living relatives but also the waxen images of all the predecessors of the deceased, -

> "imagines, quae comitarentur gentilitia funera; semperque defuncto aliquo, totus aderat familiae ejus, qui unquam fuerat, populus".[13]

From Polybius we learn that the right to keep waxen busts or images in the atrium or public apartment of a house became a mark or privilege of noble families and on this subject Ramsay and Lanciani say: —

> "It was the custom for the sons or other lineal descendants of those who held such (curule) offices to make figures with waxen faces representing their dignified ancestors and the right bestowed by such custom or usage was called 'Jus Imaginum'. These 'Imagines' or figures were usually arranged in the public apartment ('atrium') of the house occupied by the representative of the family- appropriate descriptive legends ('tituli') were attached to each - they were exhibited on all great family or gentile festivals and solemnities; and the dignity of a family and of a gens was, to a certain degree, estimated by the number which it could display. All persons who possessed one or more of these figures, that is to say, all who could number among their ancestors individuals who had held one or more Curule offices were designated by the title "Nobiles" ... These Nobiles became gradually more and more exclusive and looked with very jealous eyes upon everyone not belonging to their own class who sought to rise to eminence in the state."[14]

The method of making these images, doubtless crude at first, was greatly advanced by Lysistratus of Sicyon who was the first to make a plaster cast from the human face itself and to form an improved image by pouring molten wax into the mould, -

> "Hominis autem imaginem gypso e facie ipsa primus omnium expressit, ceraque in eam forman gypsi infusa emendare instituit".

---

12   Pliny, H.N. XXI. 49.
13   Pliny, H.N. XXXV, 2.
14   Ramsay & Lanciani, Man.rom. Antiq., p. 94.

Before the time of Lysistratus, says Pliny, modellers aimed at making, not exact likenesses, but faces as beautiful as possible, -

"ante eum quam pulcherrimas facere studebant".[15]

The art of encaustic painting in which the dry colours were blended with wax and oil and afterwards hardened by heat, was brought to a high degree of perfection by the Romans. Not only were ornaments and walls decorated in this way but, as Pliny tells us, even ships of war and cargo vessels were so painted, the molten wax being applied to their sides with brushes:

"classes pingi ... resolutis igni ceris pencillo utendi",

and these paintings, says Pliny, were not injured by sun, salt water, or winds.

The crews of warships, he adds, when about to go to death or at least to slaughter took a delight in being borne in highly decorated vessels, -

"Juvat pugnatures ad mortem aut certe caedem, speciese vehi".[16]

Referring to the wonderful durability of the Roman encaustic paintings, Cowan remarks, -

"It is certain that this method of painting resisted injury caused by time and the elements, for the vast mural paintings taken from Patrician houses in Herculaneum and Pompeii that can still be seen after the lapse of more than eighteen centuries bear witness to the wonderful preservation of these colours".[17]

As in our own times, wax was largely used for making tapers and candles, the Roman "funale" being a sort of candle made of wax or resin and provided with a woven or rush wick. Torches also were sometimes made of pure beeswax and some of these, for use on festive or ceremonial occasions, were of great size. A portion of a huge Roman wax torch discovered at Vaison in the South of France is shown in illustration No. 19. It has an external diameter of four inches and an internal portion or core, apparently consisting of a darker kind of wax, of a diameter of two inches. The wax, although very brittle, has kept its natural colour and brightness. Apparently, the wick has perished.

15	Pliny, H.N. XXXV. 44.
16	Pliny, H.N. XXXV, 31
17	Cowan, Wax Craft, p. 25.

PORTION OF ROMAN WAX TORCH

Found in Vaison near Orange in S.E. of France.

British Museum

The use of wax candles has been associated with religion since very early times and by the Romans they were specially used in the worship of Saturn, Bacchus and Ceres.[18]

The early Christians continued many pagan customs, amongst them being the use of waxen candles at their shrines and this practice has continued until the present day. Indeed, during the Middle Ages, the profitableness of beekeeping depended largely on the great demand for wax on the part of the Churches. Even today the demand is still great, for the candles used in the Roman Catholic church must be of pure beeswax only.

Morley surmises that "it was because of the supposed purity of the bee that honey and wax had their significance in religious ceremonies. Church candles must needs be made of pure unadulterated beeswax".[19]

Beeswax is composed principally of two organic bodies, myricene and cerotic acid[20] and is entirely indigestible. Notwithstanding this, however, it appears to have been

---

18   Cowan, Wax Craft. p. 81.

19   Morley, p. 324

20   Cowan, Wax Craft. p. 51.

liberally used as an internal medicament and according to Pliny it was emollient, heating and flesh-forming, -

"Omnis autem mollit, calefacit, explet corpora";

the new wax being considered best for these purposes. It was given in broth for dysentery and even the combs, mixed with baked wheaten pottage, were administered for this complaint, -

"Datur in sorbitione dysentericis, favique ipsi, in pulte alicae prius tostae".

It was believed to be contrary in properties to milk, and ten wax pills of the size of a grain of millet were supposed to prevent the coagulation of milk in the stomach, -

"non patiuntur coagulari lac in stomacho".

As an external application it was a remedy for a swelling in the groin, -

"Si inguen tumeat, albam ceram in pube fixisse remedio est".[21]

These remedies were, of course, practically useless, but beeswax still figures largely in the manufacture of various ointments and salves and according to Cowan, who gives a number of modern recipes for the use of wax, it is still used in France (in conjunction with quinces) as a remedy for dysentery.[22]

One of the most interesting purposes for which wax was needed by the Romans was that of making writing tablets ("tabellae"). These were made of thin sheets of maple or citron wood, oblong in shape, with thick raised borders, and in appearance were much like a small framed school slate. The sheet of wood inside the border was covered with a thin layer of beeswax, "Cera ... rasis infusa tabellis".[23] And on this the writing was done by means of a "Stylus" or "Graphium": an iron or bronze pencil brought to a point at one end and flattened at the other. With the point of the stylus the characters were easily cut in the wax and erasures were effected by pressing on the wax with the flattened end, whence the expression "vertere stylem", - to erase.

Perfect examples of a wax tablet and stylus are shown in Illustration No. 20, the former still showing clearly the Greek characters engraved on it nearly 1800 years ago.

21	Pliny, H.N. XXII, 55.
22	Cowan, Wax Craft, p. 147.
23	Ovid, Art. Amat. I, 437.

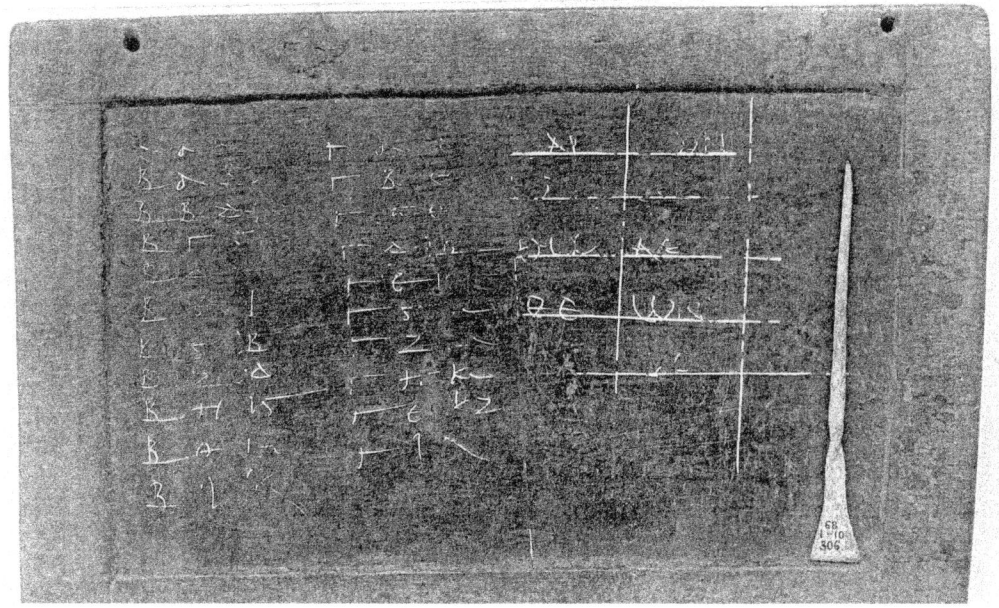

ROMAN WAX TABLET AND STYLUS

The wooden frame is pierced with two holes by means of which the tablet was tied to others to make a "Codex" or "Codicillus."

On the wax surface are the multiplication tables in Greek to 3 times 10, and a list of words divided to show stems and terminations. The Stylus is seen on the right.

British Museum

On one side the border of a tablet was pierced with holes by which it could be bound to others with flaxen string, "linum". Two or more tablets were bound together to form a book called a Codex or Codicillus. If small, such books were called "Pugillares" and were known as "Diptychi", "Tritychi", "Triplices", etc, according to the number of tablets they contained.[24] They were fastened together with linum and a waxen seal which was sometimes impressed with a signet ring. The whole of the necessary materials for writing a letter are mentioned in one line by the playwriter Plautus (born 254 B.C.), in whose "Bacchides" occurs the following dialogue, -

Chrysalus: "Nunc tu abi intro, Pistoclere, atque effer cito –"

Pistoclerus: "Quid?"

---

24   Ramsay & Lanciani, Rom. Antiq. 514-5.

Chrysalus: "Stilum, ceram, et tabellas, et linum".[25]

Tablets were commonly used for letter writing and frequently, as in the case of Ovid and Corinna, for the transmission of amatory epistles between lovers.[26] They were conveyed by the hands of slaves and the recipients, after having sufficiently read the contents, obliterated them with the flat end of the stylus and sent replies on the same wax.

Ovid, sending a missive to his mistress Corinna by the hand of her maid Nape, bids the latter take his tablets, well filled that morning, to her mistress,

"Accipe, et dominam peraratos mane tabellas Perfer …".

She is to watch the expression on her mistress' face as she reads - whether hopeful or otherwise, and if asked what he is doing, to say that he is living in the hope of seeing her that night; the flattering wax inscribed by his own hand will say the rest, -

"Caetera fert blanda cera notata manu".

She is to ask Corrina to send a long and closely-written answer, for, he says, he hates it when wide spaces on the bright waxen sheets are empty, -

"Odi, cum late splendida cera vacat".

He hopes his eyes may be long detained by her writing in the extreme margins, -

"… oculosque moretur Margine in extremo littera rasa meos",

but he adds, it will be sufficient if she will but write the one word "Come", -

"Hoc habeat scriptum tota tabella Veni."[27]

Receiving an unfavourable reply, the poet is furious. Throwing the tablets into the road where he hopes they will be broken by passing wheels, he first curses the sheets of wood,

"Ite hinc difficiles, funebria ligna",

and then the wax, bringing back the words of denial,

"Tuque negaturis cera referta notis"

and which he says must have been sent by the Corsican bee which gathered it from the tall hemlock from under its infamous honey, -

---

25   Plantus, Bacch. IV, 4. 64.
26   Ovid, Amores, I. XII, 10.
27   Ovid, Amores, I, El. XI.

"Quam, puto, de longae collectam flore cicutae

Melle sub infami Corsica misit Apis".

Its colour, which had previously appeared to him to be fair and bright, now seems to be blood red ("sanguinolentus") as though coloured with minium or red-lead, - an allusion to what was probably a common mode of colouring inferior wax and pigments. Finally, he would pray that a rotting old age may consume his tablets and that their wax may become white with foul mould, -

"Quid precor irratus? Nisi vos cariosa senectus

Rodat, et immundo cera sit alba situ".[28]

---

[28] Ovid, Amores, I, El.XII.

7

POLLEN

The subject of pollen, the fertilising "dust" derived from the anthers or male parts of flowers, is fascinating alike to the bee-man who contemplates his bees as they return to their hives laden with this rich and multi-coloured food, and to the lover of Natural History who appreciates the mutual adaptations of flowers and insects for the marriage of the former and the nourishment of the latter.

During the greater part of the year, while breeding is in progress, the bees collect great quantities of pollen. They gather it from the flowers by means of feathered hairs with which their bodies are amply provided, and by means of fine combs they collect it into two baskets or "corbiculae" which are situated on the tibiae of the hind legs.[1] After being carried to the hive, it is pressed and stored in the cells of the combs, but as it is the only source of the nitrogenous and mineral constituents of the bees' food, it is consumed almost as fast as it is brought in.

There is great diversity of colour in the beautiful balls of pollen as they are brought home by the bees, and the experienced observer can tell without difficulty from what particular kind of flowers they are gathered. Red, yellow, brown, white and black, as well as many intermediate hues are all represented, and, what is perhaps most striking, these are never mixed, for as a rule a bee keeps to a particular kind of flower during a single journey.

> "Brushed from each anther's crown, the mealy gold
>
> With morning dew, the light-fanged artists mould,
>
> Fill with the foodful load their hollowed thigh,
>
> And to their nurslings bear the rich supply."[2]

Aristotle considered that pollen ("κήρινθος") was a kind of bee-bread, and states that it was sweet like figs and carried by the bees on their legs.[3] Vergil, too, refers to the latter point when he says that the young bees return "crura thymo plenae".[4] According to Pliny,

---

1   Cowan, Honey Bee, pp. 35, 6.
2   Evans, The Bees I, 251-4.
3   Aristot. H.A. IX. 40. 2.
4   Vergil, Georg. IV, 181.

pollen was variously known as "erithace," "sandaracha" and "cerinthus" and was the food of the bees while they worked, -

"Hic erit apium dum operantur cibus."

He says it is found packed away in the empty cells, - "favorum inanitalibus sepositus," and (here he differs from Aristotle) that it is of a bitter taste, - "amari saporis".[5] It is difficult to say that pollen is either sweet or bitter. Certain kinds are a little bitter and none are properly called sweet, but in all cases it gives an objectionable flavour to honey with which it may become mixed.

It is difficult to account for the lack of observation on the part of the ancient beekeepers with regard to pollen-collecting, a phenomenon which is commonly noticed today even by school-children. Pliny, who is the only one of the ancient writers to say much about it, does not even recognise that it is gathered from most flowers. He gravely states that it is produced from the "Springdew" (rore verno) and the juice issuing from trees in the manner of gum, - "gignitur autem rore verno, et arborum suco cummium modo,"[6] from which it is apparent that he confused it with bee-gum or propolis.

He goes on to say, quite erroneously, that the amount and colour of the supplies depended on the prevailing direction of the wind, being less when the southwest ("africi") wind blew, darker ("nigrior") when the wind was in the south and better and redder when the northerly winds prevailed. The greatest quantities, he adds, were borne by almond trees ("Graecis nucibus").

It is quite true that this tree is a valuable source of both honey and pollen and, being practically the earliest fruit tree to flower (its blossoms appearing before the leaves), its pollen is extremely welcome to the bees. Its importance as a source of pollen, however, was exaggerated probably on account of the conspicuousness of the bees working on its early blooms.

After describing how he thinks the bees load their legs, Pliny tells us that three or four bees receive and unload each of the returning workers-

"excipiunt eas ternae quaternaeque atque exonerant";

but this is not a fact, for although they sometimes take nectar from the tongues of the returning foragers, they never relieve them of their loads of pollen; nor could they if they would, for once the pollen masses have become detached, they cannot be replaced in the bees' corbiculae. The loaded bee relieves herself of her burden by inserting her abdomen and hind legs into a cell and rubbing the latter together so as to dislodge the

---

5   Pliny, H.N. XI. 7.

6   Pliny, H.N. XI. 7.

pollen masses. These are then pressed down compactly and left until needed for food.

The importance of leaving a store of pollen in the hive for winter was recognised, for Pliny, speaking of the heather honey, says that two-thirds of this must be left for the bees and always those parts of the combs which contained erithace, -

"Relinque … semper eas partes favorum quae habeant erithacen."[7]

Columella appears not to have understood the nature of the stored pollen which he calls "rubras sordes" but he appreciated the fact, well-known to modern beekeepers, that it vitiates the honey on account of its bad flavour, and advises that the combs in which it is found be cut away from the rest before the honey is strained.[8]

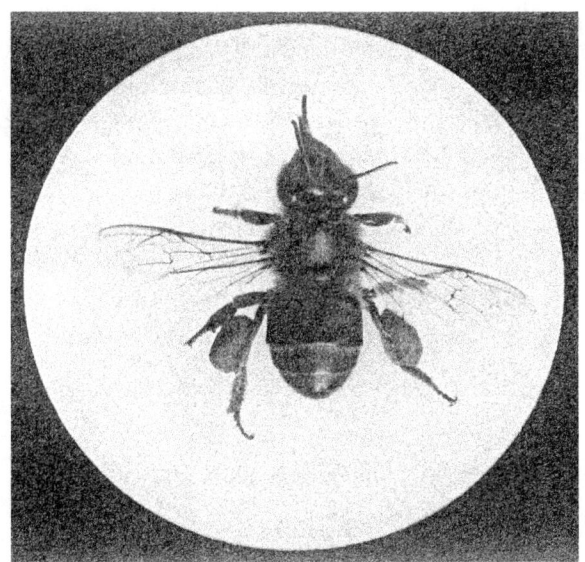

A POLLEN-LADEN BEE

"Convenitur erithace quam alique sandaracum, alii cerinthum vocant. Hic erit apium dum operantur cibus."

Pliny. H.N. XI, 7.

"Flores comportant prioribus pedibus femina onerant … totaeque onustae remeant sarcina pandatae."

Pliny. H.N. XI, 10.

---

7   Pliny, H.N. XI. 16.
8   Colum. R.R. IX, 15. 13.

8
PROPOLIS

Although the honey-bees do not build or hollow out their dwellings as do many other insects, yet Nature has provided them with an extraordinarily good building material in the form of "Propolis", a tenacious gluey substance which they gather from buds and other parts of plants and with which they repair and make snug the homes of their adoption.

Many trees, especially the poplar, willow, alder, horse chestnut and firs, provide propolis, which is a gummy exudation found in the axils of their leaves, bracts and leaf-buds. The bees gather it in threads by means of their mandibles and pack it in the pollen baskets of their hind legs. It is removed from these by other bees, and is used to stop crevices, to cement together moveable parts of the hive and, mixed with wax, to make strong attachments between the combs and the hive-roofs. It is a great nuisance to modern beekeepers for it sticks to the hands and disfigures the wooden frames and sections which contain the combs. It is carried into the hives in small quantities during the spring and summer and in larger amounts in the autumn when it becomes necessary to line the hives and stop crevices as a protection against cold.

In the following beautiful passage from Evans the gathering and use of propolis are aptly described, -

> "With merry hum the Willow'd copse they scale,
> The Fir's dark pyramid, or Poplar pale,
> Scoop from the Alder's leaf its cozy flood,
> Or strip the Chestnut's resin-coated bud,
> Skin the light tear that tips Narcissus's ray,
> Or round the Hollyhock's hoar fragrance play.
> Soon tempered to their will thro' eve's low beam
> And linked in airy bands the viscous stream,
> They waft their nut-brown loads exulting home,
> That form a fretwork for the future comb.
> Caulk every chink where rustling winds may roar,
> And seal their circling ramparts to the floor."[1]

---

1   Evans, The Bees III, 177-188.

When fresh-gathered, propolis is usually plastic and reddish brown in colour but it hardens with age and ultimately becomes almost black. It was probably on account of its changing appearance that the ancients concluded that it was not one substance but several. Thus Aristotle speaks of "Commosis" ("κόμμωσις"), with which floors were smeared and large entrances narrowed;[2] of "mitys" ("μίτυς"), the black substance spread at the entrances; and of "pissoceros" ("πισσόκηρος") smeared on the floors,[3] the first named probably including the other two.

Varro tells us that "propolis" was the name given to the substance which the bees use to form a sort of porch over the entrance of a hive, especially in the summer, -

> "propolis vocant, e quo faciunt ad foramen introitus protectum in alvum maxime aestate."[4]

Modern beekeepers who have hives with adjustable entrances seldom find them narrowed down by "bee-glue" but considerable quantities of this material are sometimes seen at the mouths of old straw skeps and it was doubtless this phenomenon which gave rise to the name "πρόπολις" – "before the city."

In another passage, however, Varro confuses propolis and pollen (erithace) for he says it is with the latter that outside the hive entrance they block up all openings by which draughts may reach the combs, -

> "Extra ostium alver obturant amnia, qua venit inter favos spiritus, quam ἐριθάκην, appelant Graeci,"[5]

and again he distinguishes erithace from propolis and says that the bees use it to join the outermost combs together, -

> "Erithace vocant quo favos extremos inter se conglutinant, quod est aliud melle, propoli."[6]

It is possible that the term erithace was sometimes loosely applied to both pollen and propolis, but it is more likely that Varro was writing inaccurately in these passages, especially as bees stop crevices only from the insides of the hives and use wax when they join their combs together.

Aristotle, Pliny and Columella use the term Erithace for bee-bread or pollen. Gesner, unable to reconcile what Varro and Pliny say about it, aptly remarks that not even the

---

2   Arist. H.A. IX. 40.3.
3   Arist. H.A. IX. 40.5.
4   Varro, R.R. III, 16-23
5   Varro, R.R. III, 16.8.
6   Varro, R.R.

skilled are always agreed with respect to the technical terms they use, -

"Scilicet ne periti quidem talium rerum eadem semper nomenclatura utuntur".[7]

Vergil says that the bees stopped up crevices with wax, -

"nequiquam in tectis certatim tenui cera

Spiramenta linunt …"

and that they filled out their borders ("oras") with "fucus and flowers," (whatever these may mean - possibly "paint" and pollen); but he is not to be taken too literally in his use of these terms for he adds that for these purposes the bees preserved a collected glue more tenacious than bird-lime or the pitch of Phrygian Ida, -

" … collectumque haec ipsa ad munera gluten

Et visco et Phrygiae servant pice lentius Ida".[8]

Pliny is more specific in his references to propolis. He tells us that the bees make "melligo" from the tears of trees which secrete glue ("quae glutinum pariunt") and gum and resin from the juice of the willow, elm and reed. With these they line the whole of their dwelling as with a kind of plaister ("quodam tectorio") having first mixed them with bitter juices, as a protection against other little creatures which are greedy of their honey, for, says Pliny, they know they are about to construct what may be greatly coveted ("quod concupisci possit"); and with these same substances they build around their wide entrances ("fores circumstruunt".[9] He then describes what he takes to be three kinds of the gummy matter:

(1) Commosin, which he says makes the first foundations ("prima fundamina"); - it is laid on as a first coating ("crusta prima") and is of a bitter taste;

(2) Pissoceros forms the second layer ("super eam venit"), a more dilute kind of wax as it were ("ceu dilutior cera"), from the softer gums of vines and poplars;

(3) Propolis, of fatter substance ("crassioris materiae") with flowers (pollen) added, not quite wax, but serving as the support of the combs ("favorum stabilimentum"); it is of a strong odour ("odore gravi") and on that account many people, says Pliny, used it instead of galbanum (a strongly scented gum used in medicine).

Pliny's differentiation of the kinds of gummy matter used by the bees is ingenious but not based on facts. So far as beekeepers know today, only one substance, commonly

---

7   Gesner, Adnott. Varro R.R. III, 16.8.
8   Vergil, Georg. IV, 38-41.
9   Pliny, H.N. XI, 6.

called propolis, is utilised, although, being collected from various sources, this may yet prove to vary considerably in composition and properties.

Propolis was used by the ancients as a remedy in several morbid affections of the body, always apparently for external application.

Aristotle tells us that "mitys" was used for bruises and suppurations. And Varro states that doctors used propolis for making plasters, -

"Quam rem etiam nomine eodem medici utuntur in emplastris,"

on which account, he adds, it was sold in the Via Sacra for more than honey.[10] To those who are familiar with the wonderful tenacity of propolis, there is no wonder that it was used for plasters, for there would be no likelihood of these coming off by accident when once applied.

According to Pliny, propolis extracts from the flesh stings and all other infixed foreign bodies, - "aculeos et omnia infixa corpori extrahit," it dissipates tumours, - "tubera discutit", and brings hard swellings to a head, - "dura concoquit"; it soothes nerve pains, - "dolores nervorum mulcet", and ulcers which are past hope it covers with a healing skin, - "hulceraque jam desperantia cicatrice includit".[11]

Propolis is not now used in medicine and is seldom collected for commercial purposes. It is sometimes utilised, however, as a constituent of certain varnishes, but the difficulty of obtaining it in considerable quantities - an ordinary beehive would contain less than an ounce - must always restrict its usefulness.

---

10   Varro, R.R. III. 16. 23.
11   Pliny, H.N. XXII. 50.

9
HONEY BEVERAGES

The essential basis of all fermented drinks is sugar, of which alcohol is a product. Any sweet substance placed in water for a few days is attacked by yeasts which convert the contained sugar mainly into alcohol and carbon dioxide. It was therefore an easy matter even for primitive men to prepare more or less delectable drinks either from the juices of fruit or from honey.

The earliest literature consequently contains frequent references to wine, and even the Homeric gods were considered to partake of the celestial foods Nectar & Ambrosia. The exact nature attributed to these cannot be ascertained, but Keightley, who reviews the references to them, concludes that Nectar was a red drink handed about in cups and Ambrosia either a similar drink or a kind of solid food. He does not mention honey as an ingredient of either, but many people have liked to believe that Nectar was honied wine (mead) or something similar, and that Ambrosia was either honey itself or a food many times more delicious.[1] Both were supposed to be elixirs of immortality and therefore forbidden to men.

But whether honey was or was not a constituent of the mythical drink of the gods, it is certain that it was largely used by the Greeks and Romans either to sweeten the fermented juice of the grape or to make the popular beverage which now goes by the name of Mead.

Although most ancient peoples are known to have valued this beverage, it was more especially the favourite drink of the early Northern peoples, and Scandinavian mythology contains numerous references to its virtues as a drink for heroes, - to be drunk liberally from golden goblets or capacious horns, not only before battle and after victory, but finally from the skulls of former enemies in Valhalla.

The brewing of mead has continued through the ages until the present day and there is very little difference in some of the methods recommended by Columella and Pliny and those followed by our Saxon forefathers and still in use today.

In order to make good mead it is necessary to mix honey and water in certain proportions. These are boiled so as to kill all undesirable yeasts or fermentative bacteria; a small amount of yeast, either natural, such as is found in all fruits (e.g. the grape), or

---

1   Morley, P. 222 & 299; Keightley. Mythology. Appendix; De Soignie, p. 114.

artificially prepared, such as that used by brewers, is introduced into the sterilised and cooled liquid and this is then allowed to stand for a few weeks in a slightly warm place until fermentation ceases. If pure honey is used, it is necessary to add small quantities of mineral salts and nitrogenous matter, for the yeast cells need these in addition to sugar in order that they may grow and multiply. If, however, honey containing pollen is used, such as is often obtained by rinsing the combs after most of the honey has been pressed from them, there is no need to add other ingredients, for the pollen grains provide the necessary mineral and nitrogenous substances. The character of the mead depends on the proportion of honey used. If this is small, as when the wet honeycombs are rinsed in water, the yeast cells continue their work until all the honey is destroyed and the result is a weak dry wine. But if more honey is added, the fermentation continues until a maximum of about 15 per cent of alcohol is obtained. At this stage the yeast ceases to work, and if there is an unchanged excess of honey the result is a fairly strong but sweet wine, which, if properly bottled, will keep and improve for years. The best results are obtained when from 3 to 4 lbs of honey are used with a gallon of water.

The Roman writers give a number of interesting recipes for making honied wines. The most esteemed of these was "Mulsum", for the making of which Columella gives the following directions, -

"Take some new must (newly pressed grape juice) from the 'lacus' (wine-

press vat), choosing the juice which has flowed out before the grapes have been too much pressed, - 'quod distillaverit antequam nimium calcetur uva'; into an 'urna' (about 2.8 gallons); of this mix 7½lbs ('pondo X') of the best honey and having diligently stirred it, put it into a stone wine bottle ('lagoena') which is to be sealed with gypsum and placed on a shelf for 31 days. At the end of this time pour it into another vessel, cover this and place it in an oven ('in furnum')".[2]

Here the wine was supposed to derive some advantage from the smoke. The necessary fermenting yeasts would be supplied by the grape-must and the result of adding honey to this was a wine of maximum alcoholic strength, as well as of decided sweetness. The sweetening was apparently appreciated in the case of some of the sour Roman wines, for Vergil says that from the better kind of bees one will get sweet honey which will tame the harsh flavour of wine, -

"Dulcia mella ... duram Bacchi domitura saporem."[3]

Palladius gives us some interesting additional details concerning the making of mulsum. The must might be used 20 days after being taken out of the "lacus", that

---

2   Colum. R.R. XII. 41.
3   Vergil, Georg. IV. 101-2.

is, when fermentation would be in rapid progress, and into this a fifth part of best unskimmed honey was to be mixed, - (about the same proportion as recommended by Columella). The whole was then to be stirred vehemently with a rooted cane, -

"agitabis ex canna radicata vehementer",

and this operation was to be repeated for 40 or preferably 50 days.

The vessel containing the fermenting liquor was to be covered with a clean linen cloth, - "mundo linteo tegas" through which the rising froth might pass; after the 50th day it was to be skimmed with a clean hand, - and then poured into and diligently sealed in gypsum-stopped vessels. All this should be done, says Palladius, in the month of October. The brew would be improved, he adds, if in the following spring it were poured off into smaller vessels to be sealed with pitch, - ("in minora et picata vascula") and these stored away in cool underground cellars, - "terrena et frigida cella," where they should be covered with soil or river sand. If carefully treated in this way they would not deteriorate with age.[4]

These directions of Palladius correspond very closely with those given to present-day makers of mead. This is brewed in the summer or autumn and racked off into jars or casks to mature for a few months. It is then bottled and stored in cool cellars, in the same way as ordinary wine, so as to protect it from variations of temperature. Indeed, to ensure a steady temperature, it would be difficult to devise a more effective method than that of burying the bottles in sand or soil as recommended by Palladius.

From Pliny[5] we learn that mulsum made with old wine (instead of must) was always the best, - "Semper Mulsum ex vetere vino utilissimum" and this because it was supposed to be more easily mixed with honey. If made with dry wine ("en austero factum"), he says, it does not load the stomach, - ("non implet stomachum") and if made with boiled honey it does not inflate it so much as is generally the case, -

"neque ex decocto melle, minusque inflat quod fere evenit,"

- a very interesting observation and probably well founded, for grape-must, sweet wine and raw honey would all contain ferments which would remain active and gas-producing in the mulsum made from them, whereas in old sour wine, they would be temporarily inactive and those in the boiled honey would have been destroyed.

This kind of mulsum too, observes Pliny, restores the appetite, -

"Appetendi quoque revocat aviditatem cibi"

---

4   Pallard. R.R. XI, Tit. XVII.
5   Pliny, H.N. XXII, 53.

for which reason, doubtless, it was used in the "Gustus" or Promulsis", the first course of a banquet.⁶

As a cold drink, continues Pliny, it moves the stomach, - "alvum mollit frigido potu", and when taken hot it usually had the reverse effect, - "pluribus calido sistet." Moreover, it causes a person to put on flesh, - "corpora auget"; above all it promotes long life and healthy old age, as in the case of Pollio Romilius who attributed his 100 years of age to mulsum within and oil without, -

"Intus mulso, foris oleo";⁷

and lastly it was considered to be a cure for the jaundice, on which account, according to Pliny, Varro called this the "royal" disease ("King's evil").

"Varro regum cognominatum morbum arquatum tradit, quoniam mulso curetur"⁸

In Pliny's time the mulsum made with wine had long superseded that made with must because the latter, called "Melliletes" induced flatulency. It was formerly given says Pliny, as a remedy for persistent costiveness ("inveteratum alvi"), for feverishness ("in febre"), for gout ("articulorum morbo"), for nerve troubles ("nervorum infirmitate") and to women who abstained from ordinary wine ("mulieribus vini abstemiis").⁹

From the writings of Columella & Palladius however it is evident that must continued to be used in the making of mulsum long after Pliny's time.

For those who could not afford the costly mulsum, the simpler and more homely mead known as "aqua mulsa" or "hydromel" was available and for this the Romans had several recipes.

As in our own time the traces of honey adhering to the pressed combs were dissolved in water. The resulting weak solution would ferment naturally and in a few days become a pleasant slightly alcoholic drink; but this was more usually boiled (the yeasts being thus killed) and left for some time exposed to air, with the result that acetic fermentation ultimately set in with the production of vinegar, -

"Vasa mellaria aut favos lavari aqua praecipiunt hac decocta fiere saluberrimum acetum".¹⁰

The previous boiling of the liquid was unnecessary and undesirable for new yeasts had

---

6   Becker, Gallus p. 380; Ramsay & Lanciani p. 495; Varro, R.R. III. 16.2.
7   Pliny, H.N. XXII, 53.
8   Pliny, H.N. XXII, 53.
9   Pliny, H.N. XXII, 54
10 Pliny, H.N. XXI, 48.

to be absorbed from the air before the honey could be changed into alcohol, a necessary preliminary to acetous fermentation. The vinegar thus produced would be weak unless honey were added to the sweetened water, but the product in any case would, as Pliny says, be very wholesome. Even today honey-vinegar is reputed to be better than that made from wine and is highly valued.

Columella tells us that the sweet liquor obtained by soaking honeycombs either in rain or spring water was boiled down to the consistency of "defrutum" (grape juice boiled to half its original volume), and that this was preserved in sealed vessels, to be used by some as a drink instead of "aqua mulsa" and by others as a preservative for olives. As a remedy for the sick, however, it was not to be substituted for "aqua mulsa" for it produced flatulency.[11]

The recipes for making mead itself as given by Pliny, Columella and Palladius are essentially sound although strange in certain particulars. Pliny, for example, tells us that some people prescribed that rainwater should be kept five years for the purpose,

- "Quinquennio ad hoc servari coelestem (aquam) jubent"

while others used that which was freshly fallen. This they boiled down to two-thirds of its original volume, afterwards adding one-third of old honey,

- "tertiam mellis veteris adjiciunt".

The mixture was then allowed to stand in the sun for forty days after the rising of the Dog Star (July 16th), although some bunged the vessels in which it was contained after ten days.[12] Palladius, in giving what is practically the same recipe, is not so particular about the choice of water, which, he says may be taken from a spring, - "ex fonte". He recommends one part of new unskimmed honey to three parts of water and that this be placed in pots in which grape juice was usually boiled down ("Carenariae") and kept stirred (and presumably boiled) for five hours. After this it was to be placed in the open for forty days.

It may be observed that the beverages made with these recipes, in which the proportion of honey was either one-third or one-fourth, would be very sweet. The former corresponds to the use of 7lbs and the latter to $4^2/_3$lbs of honey to a gallon of water. Since about 3 pounds of honey to a gallon of water are sufficient to yield 14 to 15 per cent of alcohol, after which the yeast fermentation ceases, it is obvious that while the hydromel of Palladius would be a sweet wine, that made according to Pliny's recipe would be excessively charged with unchanged honey.

11	Colum. R.R. XII. 11.
12	Pliny, H.N. XIV. 20.

Columella was aware of this for he gives two recipes, the first prescribing one pint of honey to two pints of water for a sweet wine, -

"Sive dulciorem mulsam facere volunt; duobus aquae sextariis sextarium mellis permiscent;"

and for a drier (but still sweet) wine, 9 ounces of honey to a pint of water, -

"Sive austeriorem; sextario aquae dodrantem mellis adjiciunt".[13]

Pliny remarks that the richer hydromel, if kept till old, acquired the flavour of wine and nowhere was it more commendable than in the country of Phrygia.[14]

Like mulsum, mead was considered to have valuable dietetic and medicinal properties. If freshly made it was specially good for invalids placed on light diet, - "preclarum utilitatem habet in cibo aegrotantium levi"; it was effective in restoring strength ("viribus recreandis"), and in soothing the mouth and stomach ("ore stomacho mulcendo"); and for reducing high temperature ("ardore refrigerando"). According to some authors it was useful, when taken cold, for easing the bowels ("frigidam utilius dari ventri molliendo"); it was recommended as a drink for those who feel the cold ("alsiosis") and for persons who are of feeble courage or mean-spirited, ("animi humilis et praeparci").[15]

Moreover, it was considered beneficial for those with coughs ("tussientibus utilis") and when warmed, effective in inducing vomiting ("calefacta invitat vomitiones"). With oil it was an antidote for white-lead poisoning ("venemum psimmythii") and, especially if taken with asses' milk, for poisoning by henbane and nightshade ("contra hyoscyamum .... et halicacabum"). It was poured into the ears, and used for fistulas of the generative organs, ("infunditur et auribus et genitalium fistulis"), and was used in bread-poultices ("cum pane molle") for the womb, for sudden tumours and for sprains ("vulvis, subitis tumoribus, luxatis").[16]

Notwithstanding all these supposed virtues of mead however, the later writers, says Pliny, condemned its use when old as being less hurtful than water and less strengthening than wine, - useless to the stomach and injurious to the nerves ("stomacho inutilissimum nervisque contrarium"), - which criticisms can justly be applied to much of the badly made mead of modern times.

13	Colum. R.R. XII, 12.
14	Pliny, H.N. XIV, 20.
15	Pliny, H.N. XXII, 51.
16	Pliny, H.N. XXII, 52.

An ancient liquor, of which the basis was honey, is described by Dieuches (cited by Pliny). It was called "Oxymel" and was made by repeatedly boiling together the following ingredients, -

10 Minae (about 7½ lbs) of honey, 5 heminae (about 2½ pints) of old vinegar, 1¼ lbs sea salt and 5 sextarii (pints) of sea water.

It is reassuring to read that this nasty concoction was used only as a medicine and that the physician Asclepiades condemned it and put it out of use. Before this time it had been prescribed for fevers ("in febribus"), for bites of the serpent known as "seps"; as an antidote for opium and mistletoe ("contra meconium ac viscum"); as a warm gargle for the quinzy ("anginis calidum gargarizatum"); for the ears ("auribus"); and for complaints of the mouth and throat ("oris gutturisque desideriis").[17] With reference to the last it is interesting to note that honey and vinegar, without the sea-salt, however, is still a favourite homely remedy for affections of the throat.

As might be expected, honey was used to sweeten and flavour a number of less-known beverages drunk by Romans, but these were not necessarily fermented and the honey was not, as in the cases of mulsum and hydromel, the principal ingredients in their composition.

In modern times the production of mead has largely given place to the manufacture of beer and it is now made mainly in private houses. In France, however, it is still made on a commercial scale and a strong effort is now being made to restore it to popular favour on account of its recognised superiority, when it is well made, to the "vin ordinaire" of commerce.[18]

The practice of mixing herbs and spices with honey-wines, which is quite common amongst cottagers of to-day, dates at least from Roman times. Florentinus gives several recipes for making mulsum and hydromel, amongst them being one for a spiced "oenomel" as he calls it. This was made with Attic honey and wine to which were added measured quantities myrrh, cassia, nard (leaves of the spikenard plant) and peppermint.[19] But as in our own times such admixtures were not always regarded as improvements and Florentinus himself, speaking of hydromel which was made of honey and water only, remarks that experienced physicians preferred it as they knew of what it was made, -

"Hoc etiam rerum experti medici in morbis utuntur, scientes quod ex sola aqua et melle sit compositum.[20]

17	Pliny, H.N. XXIII. 29.
18	Jungfleisch, L'Apiculteur. Mars. 1922. P. 69.
19	Geop. VIII, 15.
20	Geop. VIII, 17.

10

BEE FLOWERS

As the ancients did not properly understand the origin of honey and were unaware of the function of insects in the fertilisation of blossoms, we might expect that their writers would frequently be in error when recommending particular flowers as suitable for bees. Notwithstanding many of their misconceptions, however, it must be admitted that they were remarkably accurate in their estimation of the value of many honey plants and their lists contain the names of practically all the flowers which are highly esteemed by apiculturists of the present day.

Aristotle mentions an important general fact which does not appear to have been repeated by subsequent writers and which is not even now realised by many experienced beekeepers, viz, - that the bees gather honey from only single or cup-shaped flowers, - "φέρει δ' ἀπό πάντων ἡ μέλιττα ὅσα ἐν καλύκι ἀνθεῖ".[1]

As a rule, double flowers are without male and female parts (anthers and stigmas), these having changed into petals in the process of evolution. They are not therefore succeeded by seed and consequently, not needing the visits of insects to accomplish their fertilisation, do not secrete alluring nectar.

Aristotle realised that bees took honey from the flowers by means of their tongues,[2] although in common with other writers he did not know how it was produced in the first place. The ancients could not but observe that, whether the honey fell onto plants or not, the flowers afforded to the bees their chief pasturage and consequently the cultivation of a large variety of plants in or near apiaries was recommended.

The King of bee-plants was considered to be Thyme, of which Pliny mentions two kinds, the white and the black - ("candidum ac nigricans") - which were probably "Thymus zigis" and "Thymus serpyllum," the latter being the common wild thyme from which were gathered the celebrated honeys of Hymettus and Hybla. According to Pliny the crop of thyme was considered to indicate whether the honey harvest was going to be good or otherwise, -

"Augurium mellis est, proventum enim sperant apiarii large florescente eo".[3]

---

1   Aristotle, H.A. IX, 22.5.
2   Aristotle, H.A. IX, 22.5.
3   Pliny, H.N. XXI, 31.

Theophrastus makes a similar remark, -

> "ἀφ' οὗ καί ἡ μέλιττα λαμβάνει τό μέλι, καί τούτῳ φασίν οἱ μελιττουργοί δῆλον εἶναι πότερον εὐμελιτοῦσι ἦ οὔ".[4]

Varro tells us that the Sicilian thyme honey was so good that some people rubbed the plant in a mortar and after adding tepid water sprinkled it over all the seed plots sown for bees, -

> "Itaque quidam thymum contundunt in pila et diluunt in aqua tepida: eo conspergunt omnia seminaria consita apium causa".[5]

Pliny implies a reason for this method of sowing for he says that the seeds are so small that men sow the blossoms.[6]

As previously stated, thyme honey is not now considered so good as that gathered from certain other plants, - notably the white clover - but the wild thyme is still an important source of honey in many parts of the world. Cowan, who has arranged the commonest honey-yielding plants in three classes according to their value to the bees, places Thyme in the second class.[7]

All the flowering plants of the order "LABIATAE", to which the various "thymes" belong, yield honey, but only those with the smaller blossoms are worked by hive-bees. Several of them were regarded by the Romans as second in importance only to thyme itself, the most frequently mentioned being "apiastrum" or "balm" which Varro tells us was variously known as meliphyllon, melissophyllon or melittaena.[8]

Pliny calls it "melittaena" and says that in no other flower do the bees take more delight, -

> "Nullo enim magis gaudent flore";[9]

he adds that if the hives are rubbed with it the bees will not desert them, -

> "Melissphyllo sive melittaena si perungantur alvearia, non fugient apes".

It is interesting that many old-fashioned beekeepers still grow the common garden balm (Melissa officinalis) with which to rub their skeps before these are occupied by swarms, and Benussi-Bossi and Sartori say that the custom is still prevalent in Italy, -

---

4   Theoph. VI, 2, 3.
5   Varro, R.R. III, 16, 14.
6   Pliny, H.N. XXI, 31.
7   Cowan, List of Flowers for Bees.
8   Varro, R.R. III, 16, 10.
9   Pliny, H.N. XXI, 86.

"Colle foglie della melissa si usa strofinare le arnie nelle quali si debbono raccogliere gli sciami".[10]

Of other Labiatae, Varro commends Oscinum[11] (Ocymum Basilicum, or basil); Pliny the Cunilae[12] which include the small-leafed marjoram, pennyroyal and savory; Columella, in addition to the foregoing, rosemarinum (rosemary);[13] and Florentinus adds the salviae.[14]

All these plants yield much excellent honey in Italy to-day and Benussi-Bossi and Sartori, speaking of thyme, hyssop, marjoram and lavender, say that in the districts of Italy where these do not grow wild they are cultivated in gardens in order that they may give aroma and fragrance to the common honey, -

"per dare aroma e fragranza al nostro miele comune,"

and that they are in bloom for a long period of the year ("danno una fioritura allungata").[15]

The same writers put most of the above-mentioned flowers in the highest class of those yielding honey, the Salvia Officinalis especially being of great importance on account of its wide distribution.

It is curious that the most valued of all bee flowers, the common white or Dutch clover (Trifolium repens) appears not to be mentioned by any of the old writers, although it is a very important source of honey in Italy and neighbouring countries. Italian writers place it in the highest class of honey flowers, not only on account of the excellent quality of its honey but because it blooms almost everywhere throughout the season, -

"è spontaneo dovunque e la fioritura continua tutta la stagione".[16]

Certain of its near relatives, however, belonging to the order LEGUMINOSAE are noticed by the Latin writers. Varro[17] gives "faba" (the bean), of which several varieties are cultivated in Italy to-day and yield large quantities of honey: "pisum" (the edible pea), which however is seldom worked by hive-bees; "lens" (lentils); "medica" (probably lucerne), which in some countries affords abundant and continued bee-pasturage; "et maxime cytisum" which was possibly a trefoil but more probably the laburnum (Cytisus

---

10	L'Arte di coltivare le Api, p. 192.
11	Varro, R.R. III, 16, 13.
12	Pliny, H.N. XXI, 41.
13	Colum. R.R. IX, 4, 2.
14	Geop, XV, 2.
15	L'Arte di coltivare l'Api, p.192.
16	Benussi-Bossi e Sartori, p. 192.
17	Varro, R.R. III, 16,13.

Laburnum)[18] and which, according to Columella, was good for their health, -

"nam sunt remedio languentibus cytisi".[19]

Pliny mentions in addition "ervilia" - the wild vetch, an excellent honey plant; and also "melilotos", which was most likely our "Melilotus alba", now commonly known as "sweet clover" and which to-day is being boomed as a wonderful new honey-plant. The same writer recommends that genista should be planted about the hives, for he says, the bees are very fond of this plant, -

"Oportet ... genistas circumferi alveariis gratissimum".[20]

The beautiful trees and plants of the order ROSACEAE are very largely dependent on the hive-bees for the fertilisation of their blossoms and subsequent production of their fruits and seeds, and as a reward for the services rendered them, they offer to their little insect visitors an abundance of honey and pollen. On a warm day an orchard in blossom is alive with bees and the ancient observers could not fail to realise that the fruit blooms were highly attractive to them.

And so, at the head of his list of bee-plants Varro puts the typical flower of this order, the rose ("rosa"), doubtless having in mind the common wild dog-rose which is much frequented by bees.[21] Pliny and Columella also mention the rose as a bee-flower. The latter writer includes in his list apple and pear trees, ("pyri"), almonds ("amygdalae"), peaches ("persici") and many other kinds of fruit-bearing plants ("pomiferum pleraque"). These are so numerous that Columella hesitates to name them - ("ne singulis immorer") - for in addition to the fruit trees mentioned the Order includes, amongst many others, such well-known fruit and honey-yielding plants as the Strawberry, Raspberry, Blackberry, Cherry, Plum and Sloe, as well as the familiar but less useful Hawthorn and Blackthorn.

The CRUCIFERAE, again, comprises a large number of flowering plants whose honey affords rich store for the industrious hive-bees. In places where they are grown in large quantities, such, for example, as the areas of Eastern England where brassicae (the cabbage family) are grown for seed, large crops of excellent white honey are frequently gathered.

---

18   Bonnier, Les Noms des Fleurs, p. 265.
19   Colum. R.R. IX, 5, 6.
20   Pliny, H.N. XXI, 42.
21   Varro, III 16, 13.

Columella mentions two of the brassicae, the coleworts ("vulgares lapsanae") and the charlock ("rapistrum") which are so melliferous that they cause the waxen honeycombs to abound, -

"quae favorum ceras exuberant".[22]

Pliny observes that the bees are very fond of the flower of the mustard (Sinapis Arvensis), another of the brassicae, -

" … horum floris avidissimae sunt atque etiam sinapis"[23]

and this is quite true. Most of the brassicae are cultivated for the sake of their leaves and are not allowed to bloom, but mustard is an exception, and when it is grown by farmers its continuous sheets of yellow flowers yield rich treasure to the neighbouring bees.

Another of the Cruciferae, the common wallflower, is much favoured by bees, for it provides them with honey and pollen early in the season when few other plants are in bloom. It is probably this flower which Columella had in mind when he recommended that white lilies and the not less beautiful "leucoia" be planted in gardens for the bees.[24] An alternative and more precise translation of "leucoia" is "white violets" but these, though producing honey, are but little worked by the bees. Violets however are separately commended by Columella and also by Pliny.

Of all the trees esteemed for the nectar they produce, perhaps the most important are certain of the limes, particularly the small-leaved variety (Tilia Sylvestris). When the weather is warm, this tree, which blooms in June and July, secretes an abundance of good nectar in its blossoms and often, in addition, exudes honey-dew from its leaves.

"When joins the Linden midst her twilight gloom

Ambrosial verdure with nectareous bloom,

And oft at Summer-eve her ample round

Rings with the merry swarm's tumultuous sound."[25]

---

22   Colum. R.R. IX, 4, 5.
23   Pliny, H.N. XXI, 41.
24   Colum. R.R. IX, 4, 4.
25   Evans, The Bees, III, 608-11.

The ancient writers, however, were not in accord with regard to the virtues of the lime. Vergil speaks favourably of it, calling it "pinguem tiliam"[26] and Pliny says that some of the best honey is sucked from its leaves, -

"sorbetur optimum ... e tiliae foliis";[27]

but Columella roundly condemns the limes, for, of all trees, he says, they alone are injurious -

"At tiliae solae ex omnibus sunt nocentes."[28]

The reason for Columella's objection to limes is not recorded. On several occasions during recent years, however, it has been observed that numbers of bees working on these trees fall to the ground and are unable to rise again. The reason for this is unknown but the sickness of the bees has usually been attributed to the so-called "Isle of Wight" disease. Can it be that Columella or his informants had observed a similar trouble? If so, his condemnation of lime trees is at least intelligible although unwarranted.

Columella would have near the apiary the evergreen pine ("semper virens pinus") and the scented cedar ("odorata cedrus"),[29] but these trees are not much valued by modern beekeepers although several of the pine family afford considerable quantities of honey-dew, pollen and propolis. Speaking of the CONIFERS and in particular of the Pinus Sylvestris (Scots Pine) which is ubiquitous in Europe, Benussi-Bossi and Sartori say that these trees have little honey but in certain years give a large harvest of honey-dew, -

" ... la famiglia delle Conifere, le quali hanno poco miele, ma in certe annate danno un gran raccolto di mielata".[30]

To the Coniferae belongs the yew (Taxus baccata), despised by the ancient beekeepers because the bees were supposed to avoid it, - "taxi repudiantur".[31] The plentiful pollen borne by the flowers of the male tree is carried by the wind to the tiny waiting flowers of the female tree, and consequently the yew has no need to secrete nectar to secure the services of visiting bees.

The ivies, according to Columella, should receive a place near the apiary not on account of the quality but because of the great quantity of honey which they offer, -

---

26   Vergil, Georg. IV, 183
27   Pliny, H.N. XI, 13.
28   Columella. R.R. IX, 4, 3.
29   Colum. R.R. IX, 4, 243.
30   L'Arte de coltivare le Api, p. 190.
31   Colum. R.R. IX, 4, 3.

"Ederae quoque non propter bonitatem recipiuntur, sed quia praebent plurimum mellis."[32]

But the ivy blooms very late in the year and it is only on warm days and in favoured situations that the bees are able to collect much of the nectar which is plentifully secreted in its flowers. In Italy, its fragrant honey is considered to be a cause of dysentery amongst the bees during a long winter, -

" ... miele fragrantissimo. Ma negli inverni lunghi è causa di diarrrea alle api".[33]

Of other trees good for the bees, Vergil gives the Arbutus (strawberry tree) and the grey willows, -

" ... pascuntur et arbuta passim

Et glaucas salices ..."[34]

The male flowers (catkins) of the willows yield an abundance of early pollen and are therefore much worked by the bees.

Columella commends the acorn-bearing oaks ("glandifera robora"), which, however, instead of being esteemed by modern apiarists are considered a nuisance on account of the objectionable honey-dew which they sometimes yield; the resin-yielding turpentine ("terebinthus") and mastich ("lentiscus") trees, both of doubtful value; and the red and white jujube trees ("ziziphus"), fruiting shrubs of the buckthorn family.[35]

According to Pliny the bees do not touch the blossoms of the olive tree, -

"olivae florem ab his non attingi",[36]

but in another place the same writer considers that people are deceived who think that bees do not gather from the olives, for certain it is, he says, that where the olive is cultivated the most swarms are produced, -

"quippe olivae proventu plurima examina gigni certum est".[37]

---

32   Colum. R.R. IX. 4, 2.
33   Benussi-Bossi e Sartori, p. 194.
34   Vergil, Georg. IV, 181-2.
35   Colum. R. R. IX, 4, 3 and 6.
36   Pliny, H. N. XXI, 41.
37   Pliny, H. N. XXI, 8.

Varro, following Aristotle, believed that the bees gathered wax from the olive, -

"ex oleo arbore ceram",[38]

which of course, is impossible. The olive tree is of no value to bees.

Several familiar garden flowers are named in the lists of bee-plants given by Columella and Palladius.[39] The LILIACEAE are represented by white lilies ("candida lilia"); and the hyacinth ("coelestis nominis hyacinthus"), or as Vergil calls it, the purple ("ferrugineus") hyacinth.[40]

Of the order AMARYLLIDACEAE are mentioned the narcissus ("narcissum") and the daffodil ("asphodilum"); Columella speaks of the stem of the latter ("scapus asphodeli") but why, it is difficult to imagine.

The Crocus is one of the best of the early bee-flowers, yielding liberal quantities of nectar and pollen. It is unfortunately too scarce to be of great value, but its attractiveness for the bees is recognised by Vergil, -

"pascuntur ... passim ... crocumque rubentem,[41]

- as well as by Columella and Palladius. The former is curiously in error however when he says that the bulb of the crocus is planted in order to give colour and aroma to the honey – "qui coloret odoretque mella",[42] for in the first place crocus-honey, even if stored separately, would be too early and too meagre in quantity to be taken from the hives and therefore its colour and aroma were most likely unknown; and secondly the colour of honey in no way depends on the colour of the flowers from which it is derived.

It is not uncommon, especially in badly-cultivated gardens, to see certain of the common vegetables left in the ground until they flower and seed. The parsnip, carrot, beet and onion and the edible brassicae are examples of plants which through long and continuous selection have been changed from annuals to biennials, the wild varieties being still annuals. In their first or second year, as the case may be, they bear large heads of flowers which are eagerly worked by the bees. Columella includes in his list of plants which, as he supposed, "caused the waxen combs to abound", the wild carrot ("agrestis pastinaca") and the cultivated variety of the same name ("et ejusdem nominis edomita"). In the same category he places certain of the brassicae (already referred to) and the

---

38	Varro, R. R. III, 16, 24.
39	Colum. R. R. IX, 4, and Palladius, R. R. I, 37.
40	Vergil, Georg IV, 183.
41	Vergil, Georg, IV, 182.
42	Colum. R. R. IX, 4, 4.

flowers of the wild endive ("intubi sylvestris") and dark poppy ("nigri papaveris").[43] From the flowers of all these plants the bees obtain honey, but they were probably not sufficiently plentiful to affect materially the Roman honey harvests.

The worst honeys were believed to be obtained from the woods, - especially from the arbutus and esparto grass, and common or rustic ("villaticus") honey from vegetables ("oleribus") and manured herbs ("stercorosis herbis"). The latter point at least is fallacious for manuring makes no difference to the nectar secreted by plants. Honey produced from the flowers of the kitchen garden would be quite good, but coming from various flowers it would have no distinct characteristic and there would be very little of it. Honey from mixed sources is always inferior to that obtained exclusively from the best nectariferous crops such as thyme, clover, or fruit blossom, and the common honey which Columella considered was obtained from garden and manured plants was doubtless that collected by the bees from the multitude of different blossoms which adorned the Roman countryside.

Reference has already been made to the heathers which were rightly regarded as the source of much honey, although on account of its reputed poor quality this was not highly esteemed and was mostly left in the hives for the bees.

A few other flowering plants are mentioned but they are of little importance except perhaps the Cerintha (wax flower) which is referred to by Vergil, Columella and Pliny, the last-named writer stating that it has a hollow blossom containing the juice of honey, -

"cerinthe ... capite concavo, mellis succum habente".[44]

This flower has been variously identified but it is most probably the Cerintha major of Linnaeus, - a native of Southern Europe and known in this country as Honeywort on account of the store it offers to bees. It has a drooping cylindrical yellow flower and grows abundantly in the South of France and in Italy and Sicily.

Many of the common weeds of both arable and pasture land, though obnoxious to the farmer, are yet useful to the bees and Columella does not forget to speak of them, -

"Jam vero notae vilioris innumerabiles nascuntur herbae cultisque atque pascuis regionibus quae favorum ceras exuberant".[45]

The Roman lists of bee plants were considerable, yet very many were necessarily omitted, for as Columella observes, their number was incalculable ("inexputabilis erat numerus"). Even to-day it would be an extremely difficult task to compile a complete list of bee-flowers for any particular country, for not only are almost all single flowers useful

---

43   Colum. R. R. IX, 4, 5.
44   Pliny, H.N. XXI, 41.
45   Colum. R.R. IX, 4, 5.

in varying degrees for providing nectar or pollen, or both, but so much does the secretion of nectar depend on weather conditions that a particular plant might be absolutely neglected by bees in one year and thronged with them in the next.

During the later summer months large quantities of ripe fruits are ruined by wasps which are able to pierce the skins with their mandibles so as to feed on the sweet juices within. Bees are often blamed for this for in times of scarcity they will share in the plunder although they are incapable of injuring the fruits in the first place, - a fact not generally known. It is interesting to note however that Pliny was aware of it for in justice to the bee he remarks

> "fructibus nullis noceter".[46]

---

46   Pliny, H.N. XI, 8.

11
CURIOUS BELIEFS

In all countries and in all ages many superstitious beliefs and practices have been associated with bees, and even in our own enlightened time it is quite common to meet people who scruple to purchase a swarm with money, lest this be followed by bad fortune; who would never omit ceremoniously to "tell the bees" of the death of their master; or who are quite convinced that a swarm settling on a house presages a fire and so on.

Several of the mythical beliefs relating to bees which were entertained by the Romans have already been referred to in the introduction to this paper. Certain others founded on popular ignorance and misapprehension as well as on myth, are nevertheless interesting and deserve some attention.

Foremost among these was the widespread conviction that bees could be generated from the putrefying carcases of oxen. Most of the old apicultural writers refer to this and Vergil and Florentinus give detailed directions as to the preparation of the dead animals and the formalities to be observed in order that successful results may be obtained.

According to Varro bees were born partly from bees and partly from the body of an ox which had undergone putrefaction, -

"Primum apes nascuntur partim ex apibus, partim ox bubulo corpore putrefacto".

He goes on to quote an earlier writer, Archelaus, who in an epigram had described the bees as the roaming children of a dead ox, -

"βοός φθιμένης πεπλανημένα τέκνα".

The same writer had said that wasps came from horses, -

"ἵππων μέν σφῆκες γενεά";

And bees from heifers, -

"μόσχων δέ μέλισσαι."[1]

Pliny expresses similar ideas and accounts for them by saying that the changes are due to natural metamorphoses by which some creatures are derived from others. It was believed, he says, that if bees were entirely lost they might be created anew from fresh paunches

---

1   Varro, R.R. III 16, 4.

of cattle buried in dung, -

> "Putent ... in totum vero amissas reparari ventribus bubulis recentibus cum fimo obrutis",

or from the lifeless bodies of steers, - just as wasps and hornets come from the bodies of horses and cockchafers from those of asses, -

> "juvencorum corpore exanimato sicut equorum vespas atque crabrones sicut asinorum scarabaeos".[2]

The superior size and formidable sting of the hornet receive an implied recognition from Ovid who considers that this insect has its origin in a buried war-horse, -

> "Pressus humo bellator equus crabronis origo est".[3]

Relating the story of Aristeus who sought the unwilling advice of Proteus as to how to repair the loss of his bees, Ovid makes the latter say,

> "Obrue mactati corpus tellure juvenci:
>
> Quod petis a nobis obrutus ille dabit.
>
> Jussa facit pastor. Fervent examina putri
>
> De bove mille animas una necata dedit."[4]

Vergil tells us that the practice of producing bees from the blood of slaughtered bullocks was considered infallible in Egypt. Osten-Saken, citing Antigonus Carystius, explains that the Egyptians "used to bury the ox with projecting horns, through which, after they were cut off, bees would emerge".[5] The origin of this method of bee-making is ascribed by Vergil to Aristaeus the son of Apollo and Cyrene, who first taught men the management of bees and the use of olives. The story of how he repaired the loss of his bees is beautifully related by Vergil in the latter part of his fourth Georgic. Having lost his stocks through disease and starvation, Aristaeus complains to his mother Cyrene who instructs him to go to Proteus, a sea-god who possessed the power of prophecy, which, however, he used only under compulsion. By strategy Aristaeus was to chain Proteus while asleep and force him to tell the reason of his ill-fortune. Captured and chained by stratagem, Proteus informs Aristaeus that his misfortunes are sent upon him by Orpheus whom he had once despoiled of his queen Eurydice. To this he adds that the nymphs with whom Eurydice once danced had brought about the death of Aristaeus' bees and

---

2   Pliny, H.N. XI 20.
3   Ovid, Metam XV, 368
4   Ovid, Fasti, 1, 376-80.
5   Osten Sacken.

that if he wished to obtain others, these nymphs must be propitiated. For this purpose he must select four choice bulls and as many heifers which had not been subjected to the yoke, -

"Quattuor eximios praestanti corpore tauros ...

Delige, et intacte totidem cervice juvencas;"

for these he was to erect altars at the shrines of the nymphs, and there to shed the sacred blood from their throats and leave their carcases in a shady grove. On the ninth day afterwards he was to present Lethaean poppies as funeral gifts to Orpheus, slay a black sheep, revisit the grove, and sacrifice a calf to the appeased Eurydice.

Aristaeus hastens to carry out his mother's precepts and visiting the grove on the ninth day he beholds a wondrous prodigy, - bees hum throughout the dissolved entrails and burst from the ruptured sides of the cattle, whence they trail in immense clouds, and gathering together on a tree-top suspend their cluster from the yielding branches, -

" ... liquefacta boum per viscera toto

Stridere apes utero, et ruptis effervere costis

Immensasque trahi nubes, jamque arbore summa

Confluere et lentis uvam demittere ramis".[6]

Aristaeus must have placed an excessively high value on his bees. How many farmers of today would be willing to sacrifice eight choice cattle, a calf and a sheep in order to obtain a single swarm?

---

6   Vergil, Georg. IV, 555-8.

"OXEN-BORN BEES."

From an illustration in Dryden's Vergil. 1698.

"Hic vero subitum ac dictu mirabile monstrum

Adspiciunt, liquefacta boum per viscera toto

Stridere apes utero et ruptis effervere costis,

Immensaeque trahi nubes, jamque arbore summa

Confluere et lentis uvam demittere ramis."

Vergil, Georg. IV, 554-8.

Vergil however also prescribes a less extravagant method in which a single bullock was sacrificed and allowed to putrefy in a specially constructed hut.[7] This method was described in greater detail by Florentinus, who probably lived in the 3rd Century A.D. After stating that Juba, King of Libya, produced bees in a wooden box, he goes on to describe what he calls the better method, previously recorded by two Roman writers, Democritus and Varro. The details of the *modus operandi* are absurd and revolting; -

A house was to be built ten cubits in length, breadth and height, to be provided with one door and to have a window in each of the four walls. Into this building was to be led a bull of thirty months, fleshy and very fat. He was to be killed by a number of young men who were to beat him to death with clubs in such a way as to mangle the bones without the shedding of any blood. When dead, the bull was to be turned on his back and all the orifices of the body such as eyes, nose, ears, etc., stopped with clean fine linen dipped in pitch. The carcase was then covered with thyme, and the door and windows of the house closed and sealed with thick mud so as to prevent the ingress of air. Three weeks later, light and fresh air were to be admitted, - not, however, from the windward side, - after which the house was again sealed up as before. Eleven days later the place would be found full of bees clustering together in bunches, and nothing would be left of the bull but the horns, bones and hair.

> "Undecima deinde post hanc diem ubi aperueris, invenies plenam apibus racematim inter se coagmentatis, et ex bove cornua reliqua et ossa, ac pilos, sed praeterea nihil".[8]

They say, continues Florentinus, that the ordinary bees are born from the flesh, but that the "Kings" come from the brain and spinal cord of the dead animal.

It is superfluous here to state that in no circumstances are bees generated in the dead bodies of animals, or to discuss the preposterous and impossible directions given by Florentinus. Yet the belief that bees could be so obtained was very widespread and persistent. Edwardes, speaking of Vergil's "malodorous experiment" says:

> "Not only was this practice a recognised and established thing in Vergil's time but entire credence was placed in it throughout the Middle Ages, down, in fact, to so late a time as the seventeenth century. It is on record that the experiment was carried through with complete success by a certain Mr Carew, of Anthony, in Cornwall, at an even later date still."[9]

Much has been written about Vergil's oxen-born bees and some of the commentators

---

[7] Vergil, Georg. IV, 295-314
[8] Geop. XV, 2.
[9] Edwardes, Lore, p. 8.

have accepted the view that such bees were actually produced and have attempted to account for the phenomenon. Martyn thinks it is comparable to Samson's discovery of bees and honey in the carcase of a lion, and Keightly considers that a queen bee might gain access to a carcase and deposit eggs therein.

Osten-Saken has written a long paper on the subject in which he maintains that the supposed honey-bees of the ancients were really drone-flies (Eristalis tenax). This explanation is the one commonly accepted to-day, amongst the writers subscribing to it being Lydekker, Edwardes and Royds, but to the present writer it appears to be an extremely doubtful one.

The drone-fly is a member of the large family of Syrphidae or Hover-flies, some of which have carnivorous larvae which feed on Aphidae. The drone-fly itself is noted for its close resemblance to the honey-bee, from which, however, it is readily distinguished on account of its having only one pair of wings and no sting. Like the bee it frequents flowers and feeds on honey and pollen, but on the other hand it is a solitary and not a social insect. It deposits its eggs near to the edges of dirty pools of water, so that the resulting larvae may live on the filth contained therein. Although the larva lives in the water it needs air and to obtain this it is provided with a flexible and telescopic tail traversed by tracheal tubes opening at the tip. It is able to breathe when submerged by keeping the tip of this tail above the surface of the water.

Although certain writers, including Kirby and Lydekker, state that the drone-fly sometimes lays its eggs on carcases, they do not account for the subsequent larval stage, and it is difficult to believe that a larva so well adapted for a life in water and needing air could develop and breathe inside a dead body of which the skin was not even broken. Moreover, one wonders how the parent fly, which is extremely nervous, gained admittance to the buildings described by Vergil and Florentinus before the doors and windows were sealed up.

To maintain his thesis Osten-Saken has to postulate that "during the second stage of putrescence a pool of corrupt water is formed about the carcase",[10] but this, so far as the present writer is able to ascertain, is not a fact, especially in dry situations and climates; he has to admit, moreover, that "it is a very rare now to come across a carcase and to see Eristalis tenax hovering about it"; and to assume that "in ancient times it had to look out for stray carcases. Civilisation offers it its drains, canalisations, cesspools and dung-heaps, in which it can wallow in abundance".[11]

---

10   Osten-Saken, p. 41.
11   Osten-Saken, p. 32.

Curious Beliefs

THE DRONE-FLY

(*Eristalis tenax*)

Even if drone-flies were produced in the experiment described by Vergil and Florentinus, however, they certainly would not gather in clusters after the manner of bees as is related by both writers. Osten-Saken brings scarcely any positive evidence in support of his theory and his oxen-born drone-flies are almost as improbable as the honey bees derived from the dead cattle of the ancients. It is common knowledge however, that decaying carcases are rapidly attacked by various flies whose larvae are carniverous, such for example, as the Sarcophagae, Calliphorae and Lucilae. The larvae or maggots of these flies might easily be mistaken by casual observers for the larvae seen in the combs of the bees and it was doubtless this fact and this only, which gave rise to the erroneous belief that bees were produced from decaying bodies. Under the Roman system of beekeeping, swarms and casts must have been very plentiful and their unexpected appearance in the neighbourhood of a carcase set to putrefy would be quite enough to confirm the belief that the latter was the origin of the bees. In this connection it is significant that Vergil states that the victim steer was to be slaughtered when the earliest zephyrs disturbed the waves, before the meadows blushed with new colours, and the prattling martins built their nests under the eaves,[12] - that is to say in the spring, so that the supposed bees would be produced about the time when the swarms issued from the hives.

Osten-Saken even considers that the bees found by Samson in the carcase of the lion were really drone-flies and that the "finding and eating of the honey was the myth grown out of the misconceived fact", - an entirely unnecessary perversion of what is quite a feasible story. Some time had elapsed after Samson had slain the lion before he saw it again, and in the meanwhile, the carcase had doubtless become a skeleton, thoroughly cleaned by vultures and therefore inoffensive to bees. Such a skeleton would make an acceptable home for a vagrant swarm notwithstanding Osten-Saken's trivial objection that it would require the expenditure of more wax than usual. He does not appear to have been aware that swarms sometimes build their combs in trees or hedges without any surrounding dwelling at all. Indeed, this writer's far-fetched conclusions would not deserve so much attention but for the fact that they have been widely accepted by others.

To the Romans, as to later peoples, a swarm of bees in an unusual place presaged important events. If it appeared in a camp on the eve of a battle it was believed to be an augury of evil, -

"dirum id ostentum existumant semper".[13]

Thus Livy relates that a swarm settled on the tree which overshadowed the tent of Scipio on one of the occasions when he was defeated by Hannibal, and Appien tells of another swarm which alighted on an altar in the camp of Pompey on the eve of his defeat at Pharsalia.[14] But Pliny tells us that these auguries were not always reliable, for a swarm alighted in the camp of the general Drusus when he gained his great victory at Arbalo.[15]

---

12   Vergil, Georg. IV, 305-8.
13   Pliny, H.N. XI, 17.
14   Cited by De Soignie, p. 150.
15   Pliny, H.N. XI, 17.

According to Plutarch a swarm of bees settled on the stern of Dion's ship when he set out against the tyrant Dionysius and this was considered to portend that Dion's project would flourish for a while but would afterwards come to nothing.[16] The same writer relates that a swarm of bees covered the first standard in the army of Brutus on the eve of his disaster at Philippi.[17]

But although swarms were portents of evil to warriors, they augured good for children, and especially the gift of eloquence. Pliny tells us that bees settled on the lips of Plato while he was yet an infant, thus presaging the sweet eloquence for which he was afterwards distinguished.[18] Pausanius relates a similar occurrence in the childhood of the lyric poet Pindar,[19] and Homer was believed to have derived his honied eloquence from having received his first nourishment from the honey-yielding breasts of a priestess.[20]

That the wonderful and mysterious bees should be associated with the miraculous was but natural, for as Vergil writes, they were said to possess a share of the divine mind and ethereal breath. Their souls at dissolution are restored to God, - there being no place for death; but mounting to the lofty heaven, they soar alive to their place amongst the stars, -

" … quidam …

Esse apibus partem divinae mentis et haustus

Aetherios dixere …

Scilicet huc reddi deinde ac resoluta referri

Omnia nec morti esse locum, sed viva volare

Sideris in numerum atque alto succedere caelo."[21]

---

16   Plutarch, Dion, p.665.
17   Plutarch, Lives, p.687.
18   Pliny, H.N. XI, 17.
19   Pausan. IX, 23.
20   Morley, The Honey Makers, p. 287.
21   Vergil, Georg. IV, 219-227.

12

DE APIBUS - DIDYMUS

I cannot better close this imperfect account of Roman bee lore than by translating the charming appreciation of bees written by Didymus;[1] -

"Bees are the wisest and most skilled of all animals, and in their intelligence they almost approach man; their work also is divine and to man most useful.

The government of these animals, too, is like the best regimes of organised cities. For they go out in deference to the wish of their leader and at his command. From the flowers and trees they carry the most glutinous tears, with which, as with a kind of unguent, they smear the floors and portals of their dwellings. Thence some elaborate honey, - some of the other things that they make.

They are, moreover, the cleanest of animals, alighting on no evil-smelling or unclean thing. They are not gluttonous, nor on flesh or blood or fat of any kind do they settle, but only on those things which have sweet juice.

Nor, in sooth, do they destroy the labours of others, yet most stoutly do they resist those who approach to destroy their own. Being conscious of their feebleness they make narrow and sinuous approaches to their dwellings so that if any animal shall have gained admittance in order to do them injury, they crowd around him and easily put him to death.

Yet are these little creatures soothed by sweet melody; whence, when they are dispersed, the bee-men discreetly gather them together by means of beaten cymbals or the pleasant clapping of their hands.

Alone these animals seek a leader, who may concern himself with the care of them all. On which account they always cherish their King and follow him eagerly in whatever circumstances he may be placed; they support him when he is weary and when he cannot fly, they carry and so save him.

Especially, too, do they hate sloth, for which reason the lazy ones and those given to ease and those who consume the hard-earned honey, with one mind do they put to death.

Moreover, in that they make six-angled cells their skill is seen to approach almost to human intelligence."

---

1   Geop. XV, 3.

(Note. I have kept to the plural throughout this passage. L.E.S.)

AN OLD EXMOOR BEEKEEPER

A picturesque example of a fast disappearing type of rustic beekeeper.

His methods are entirely traditional, simple in the extreme and not nearly so advanced as those described by Columella.

In his lonely cottage on the moor he lives, like Vergil's old Corycian, a life of well-requited toil and sweet content, happy in the possession of his fruitful land and friendly bees.

12
BIBLIOGRAPHY

Athenaeus.	Deipnosophistrarum, Libri Quindecim, Ed.	
		Casaboni. Lugduni, 1657.
Becker.	Gallus oder Römische Scenen aus der	
	Zeit Augusts. Metcalfe's English Translation.	London, 1844.
Beloe.	Herodotus. English Translation.	London, 1812.
Benussi Bossi	L'Arte di coltivare le Api. e L. Sartori	Milano, 1901.
Billiard.	Notes sur L'Abeille et L'Apiculture dans L'Antquité.	Lille, 1900.
Bonnier.	Plantes médicinales, Plantes mellifères, Plantes utiles et nuisibles.	
		Paris, 1921.
"	Les Nectaires.	Paris, 1879.
Bradley.	A General Treatise on Husbandry and Gardening.	London, 1720.
Brice & Campbell.	The Seven Books of Arnobius.	
	(English Translation).	Edinburgh, 1882.
Carpenter.	Bees in Palestine and Cyprus.	B. B. Journal, January, 1922.
Carr.	Early History of Bees and Honey.	Salford, 1880.
Cheshire.	Bees and Bee-keeping.	London, 1886.
Connington.	P. Vergili Maronis Opera.	London, 1865.
Cowan.	The British Beekeeper's Guide Book.	London, 1913.
"	The Honey Bee	London, 1904.
"	List of Flowers sought after by Bees.	
"	Wax Craft.	London, 1908.
Creswell.	Aristotle's History of Animals.	
	English Translation.	London, 1907.

Deane.	A note on this alleged poisonous properties of Honey from Datura Stramonium. (Including notes on honey from Azalea Pontica and Belladonna)	
	British Bee Journal, Vol. XLI. P.358.	
Dennler.	La Toilette printanière de nos ruches. L'Apiculture,	
		Novembre, 1920. Paris.
De Soignie.	L'Abeille à travers les ages.	Bruxelles, 1896.
Dryden.	Works of Vergil. (Illustrations).	London, 1698.
Du Vallius.	Aristotle. Omnia Opera.	Paris, 1654.
Edwards.	The Lore of the Honey Bee.	London, 1908.
Evans.	The Bees; a poem, in four books.	Shrewsbury, 1806.
Forbiger.	P. Virgilii Maronis Opera.	Lipsiae, 1852.
Gesner. C.	Claudii Aeliani Opera Omnia.	Tiguri, 1556.
Gesner. M.	Scriptores Rei Rusticae.	Lipsiae. 1735.
Gusman.	Pompei. The city, its Life and Art.	
	Trans. By Simmonds and Jourdain.	London, 1900.
Harduinus.	Caii Plinii Secundi Historiae Naturalis Libri XXXVII.	
		Biponti. 1783.
Heinsius.	Theophrasti Eresii Graece et Latine Opera Omnia.	
		Lugdini, 1713.
Imms.	Report on a Disease of Bees in the Isle of Wight.	
	Journal of the B. of Agriculture.	London, 1907.
Jungfleisch.	Fabrication simplifiée de l'Hydromel.	
	L'Apiculteur, Mars,	1922. Paris.
Keightly.	Notes on the Bucolics and Georgics of Vergil, including a Flora Vergiliana.	London, 1846.
Keller.	Elenchus.librorum de apium cultura	Milan, 1881.
Kirby.	Text Book of Entomology.	London, 1892.
Langhorne.	Plutarch's Lives.	London, 1873.

Langstroth.	The Honey Bee.	Hamilton, U.S.A. 1911.
Lubbock.	Ants, Bees, and Wasps.	
"	Senses of Animals.	London, 1891.
Lydekker.	The Royal Natural History.	London, 1896.
"M"	Ancient Bee Literature.	British Bee Journal, 1916.
Mair.	Hesiod. The Poems and Fragments done into English Prose.	Oxford, 1908.
Martin.	Pub. Virgilio Maronis Georgicorum Libri Quatuor.	London, 1741.
Sharp.	Insects, Cambridge Natural History.	London, 1909.
Shuckard.	British Bees.	London, 1886.
Simon.	Le Governement Admirable; ou La République des Abeilles.	Paris, 1758.
Sillig. J.	C. Plini Secundi Naturalis Historiae Libri	Hamburg et Gothae, 1851.
Snodgrass.	The Anatomy of the Honey Bee.	Washington, 1910.
Smith, and others.	A Dictionary of Greek and Roman Antiquities.	London, 1890.
Spelman & others	The Whole Works of Zenophon	London 1849
Walker.	Corpus Poetarum Latinorum.	Londini, 1849.
Wilkinson.	The Egyptians in the time of the Pharaohs.	London, 1857.

www.ingramcontent.com/pod-product-compliance
Lightning Source LLC
Chambersburg PA
CBHW081204170426
43197CB00018B/2919